SpringerBriefs in Education

For further volumes:
http://www.springer.com/series/8914

Amy Cutter-Mackenzie
Susan Edwards · Deborah Moore
Wendy Boyd

Young Children's Play and Environmental Education in Early Childhood Education

 Springer

Amy Cutter-Mackenzie
School of Education
Southern Cross University
Coolangatta, QLD
Australia

Wendy Boyd
School of Education
Southern Cross University
Lismore, NSW
Australia

Susan Edwards
Deborah Moore
Faculty of Education
Australian Catholic University
Fitzroy, Melbourne, VIC
Australia

ISSN 2211-1921   ISSN 2211-193X (electronic)
ISBN 978-3-319-03739-4   ISBN 978-3-319-03740-0 (eBook)
DOI 10.1007/978-3-319-03740-0
Springer Cham Heidelberg New York Dordrecht London

Library of Congress Control Number: 2013958138

Printed on acid-free paper

Springer is part of Springer Science+Business Media (www.springer.com)

Acknowledgments

The work presented in this book was supported by funding from the Australian Research Council under the Discovery Projects Scheme (2010–2012). Fieldwork for this project was conducted in Victoria, Australia with ethical approval from the Department of Education and Early Childhood Development and the Monash University Human Research Ethics Committee. The authors wish to thank the participating children, families and teachers for their contribution to the project. The authors also wish to acknowledge support with fieldwork and data management provided by Deb Moore, Tracy Young and Tiffany Cutter.

Material presented in this book is informed by previously published work, including:

- Cutter-Mackenzie, A., Edwards, S., & Widdop Quinton, H. (in press). Child-Framed Video Research Methodologies: Issues, Possibilities and Challenges for Researching with Children. *Children's Geographies.*
- Edwards, S., & Cutter-Mackenzie, A. (2013). 'Next time we can be penguins': expanding the concept of 'learning play' to support learning and teaching about sustainability in early childhood education. Chapter in O. Lillemyr, S. Dockett., & B. Perry (Eds.), *International perspectives on play and early childhood education*. New York: Information Age Publishing.
- Cutter-Mackenzie, A., & Edwards, S. (2013). Towards a model for early childhood environmental education: Foregrounding, developing and connecting knowledge through play-based learning. *The Journal of Environmental Education, 44*(3), 195–213.
- Edwards, S. & Cutter-Mackenzie, A. (2013). Pedagogical play-types: what do they suggest for learning about sustainability in early childhood education? *International Journal of Early Childhood.*
- Edwards, S., Moore, D., & Cutter-Mackenzie, A. (2012). Beyond 'killing, screaming and being scared of insects': learning and teaching about biodiversity in early childhood education. *EC Folio, 16*(2), 12–16.
- Edwards, S., & Cutter-Mackenzie, A. (2011). Environmentalising early childhood education curriculum through pedagogies of play. *Australasian Journal of Early Childhood, 36*(1), 51–59.

Contents

Figures

Tables

Chapter 1
A Challenge for Early Childhood Environmental Education?

Abstract This chapter orients the reader by introducing the underlying premise of the book, in addition to outlining the remaining six chapters. The book's foundation lies squarely in an era in which environmental education has been described as one of the most pressing educational concerns of our time, leading to the critical need for further insights in understanding how best to approach the learning and teaching of environmental education in early childhood education. In this chapter and indeed this book more broadly we address this concern by identifying two principles for applying play-based learning in early childhood environmental education. The principles we identify are the result of research conducted with teachers and children using three different types of play-based learning, namely open-ended play, modelled-play and purposefully framed play. Such play types connect with the historical use of play-based learning in early childhood education as a basis for pedagogy.

1.1 Introduction

Four children assembled around a wading pool at a preschool are intently engaged in play. Samples collected during a recent excursion to their local beach are the focus of attention. Seaweeds and sponges have been combined with plastic sea animals and placed in the wading pool with a small amount of water to help the children learn aspects of biodiversity. The children introduce well-known characters from a Nickelodeon™ cartoon, and one of the sponges becomes SpongeBob Squarepants™, whilst a plastic sea star morphs into his sidekick Patrick. Seaweed is heaped upon both SpongeBob and Patrick by two of the children. The remaining children swirl the water vigorously with sticks gathered from a nearby tree, creating whirlpools that lift seaweed from SpongeBob and Patrick. Seth (the teacher) observed the children at play, noting the appropriateness of their social interactions and the sophisticated articulation of the cartoon genre to their SpongeBob SquarePants dramatisation.

A. Cutter-Mackenzie et al., *Young Children's Play and Environmental Education in Early Childhood Education*, SpringerBriefs in Education, DOI: 10.1007/978-3-319-03740-0_1, © The Author(s) 2014

This typical approach to science education in the early years (Howitt et al. 2011) seems to capture what one might think early childhood environmental education should entail. Four young children are happily engaged in playing outside and experiencing nature whilst participating in an open-ended play-based activity. Elements of traditionally valued early childhood education are evident, including access to outdoor play, opportunities for freely experimenting with materials and participation in pretend play. Many early childhood teachers may have considered not much more would be needed to help these children acquire the conceptual underpinnings of the biodiversity to be found on their local beach. They may have also hoped that these children would learn to respect other living creatures and possibly embrace 'biophilic' dispositions or express an 'affinity with nature' rather than being 'biophobic' or 'afraid of nature' (Wilson 1992; Orr 1992). Now, and later, as they grow into adulthood these four children might also carry such knowledge and attitudes into their understandings about the environment and develop a commitment to living a sustainable life.

Children developing conceptual knowledge about biodiversity, understanding the importance of sustainability, and avoiding or disrupting the development of biophobic attitudes towards nature are laudable outcomes for early childhood education to achieve. However, recent research has expressed concern regarding the extent to which such exploratory play facilitates children's conceptual learning (Fleer 2010) in environmental education. In this regard, outdoor play alone has been labelled by some researchers (Davis 2010; Waller et al. 2010) as insufficient for supporting children's developing environmental attitudes and dispositions towards sustainability (Davis 2010). Experiences such as those described in the opening vignette may no longer be enough to ensure young children are having meaningful engagements with environmental education in early childhood settings. In an era in which environmental education has been described as one of the most pressing educational concerns of our time (UNESCO-UNEP 1976), further insights are needed to understand how best to approach the learning and teaching of environmental education in early childhood education. In the context of early childhood education, this is a particularly interesting concern, because the question of 'how' to approach the learning and teaching of environmental education necessarily relates to the use of play-based learning as a basis for pedagogy.

1.2 Play-Based Learning in Early Childhood Environmental Education

In this book we address this concern by identifying two principles for using play-based learning in early childhood environmental education. The principles we identify are the result of research conducted with teachers and children using different types of play-based learning whilst engaged in environmental education. The play-types used in the research connect with the historical use of play-based learning in early childhood education as a basis for pedagogy. This history is examined in Chap. 2 of the book. The principles are also informed by consideration of the environmental education literature. Here, we consider the extent to which environmental education expresses different epistemological and ontological perspectives regarding the purpose of environmental learning. In Chap. 3 we reflect on these differing viewpoints and canvass how they have been expressed to date in early childhood education, including the well-known 'Education for Sustainability' (EfS) approach.

Chapters 4–6 of the book are dedicated to showcasing the pedagogical work of teachers and children using different types of play-based learning to engage in environmental education. The final chapter, Chap. 7, reflects on this work and articulates two guiding principles for using play-based learning in early childhood environmental education. These principles are:

1. Valuing different play-types according to their pedagogical potential for engaging with aspects of environmental education; and
2. Creating combinations of play-types that support engagement with different aspects of environmental education.

The pedagogical work showcased in Chaps. 4–6 is derived from a research project conducted in Victoria, Australia over a 2-year period. The focus of the project was on examining approaches to play-based learning and how these relate to environmental education in the early years. The project involved sixteen early childhood educators and 114 children. The children were all aged 3–5 years and were attending early childhood education in formal prior-to-school settings. All sixteen of the participating educators held university level qualifications in early childhood education at the Bachelor degree (4 years) or higher level. The early childhood settings included a mixture of inner city and suburban locations, as well as a range of socio-economic levels. All settings included a culturally diverse mixture of children and families. In this book we share the pedagogical work and subsequent reflections of three of the participating teachers, including Jeanette, Josh and Robyn. At the start of each chapter we introduce the teachers and provide some background information about their interests in environmental education, their teaching and learning philosophies, and the social and educational context associated with their centres. Each chapter concludes with a brief summary of the approach undertaken by Jeanette, Josh and Robyn which highlights how the two principles of play-based learning were used in their work with the children.

1.3 Project Overview

During the course of the project teachers were invited to a professional learning session and provided with a summarised history of the role of play-based learning in early childhood education (Chap. 2). The teachers were then introduced to three 'types' of pedagogical play-based learning evident in the literature including those described by Trawick-Smith (2012) as the 'trust in play', 'facilitate play' and 'learn and teach play' approaches. The three play-types also included reference to Wood and Attfield's (2005) work describing pedagogical play along a continuum of activity in which children's self-selected play is located towards the left of the continuum and adult-framed activity towards the right (see Chap. 3). In previous work we aligned these play-types with the teaching and learning of biodiversity with young children in early childhood settings (Edwards et al. 2010), describing them as 'open-ended play', 'modelled play' and 'purposefully-framed play'. We selected biodiversity as an important environmental education concept for engaging teachers and children via these play-types because it focuses children's attention on the natural world, exposes children to opportunities for thinking about habitat, and promotes opportunities for learning to respect other living beings (Edwards and Cutter-Mackenzie 2013; Shaffer et al. 2009). The three play-types the teachers used in their engagements with the children therefore included explicit reference to biodiversity as an environmental education concept:

1. *Open-ended play*: located towards the left of the continuum and involving play experiences where the teacher provides children with materials suggestive of a biodiversity concept, and with minimal engagement and interaction allows them to examine and explore the materials as a basis for learning about the concept.
2. *Modelled-play*: located in the middle of the continuum and involving play experiences where the teacher illustrates, explains and/or demonstrates the use of materials suggestive of a biodiversity concept prior to allowing children to use the materials with minimal adult interaction as basis for learning about the concept.
3. *Purposefully-framed play*: located across the entire spectrum of the continuum and involving play experiences in which the teacher provides children with materials suggestive of a biodiversity concept and provides opportunities for open-ended play, followed by modelled-play and then teacher-child interaction/engagement.

The teachers were invited to use these three play-types to engage children in learning about biodiversity in their own settings. Each teacher was provided with a concept map outlining what topics might be associated with biodiversity as an environmental education concept. This included animal habitats, habitat destruction and plant life as relevant areas of investigation (Fig. 1.1).

The play-types were clustered into several different iterations so that of the 16 participating teachers there were small groups of teachers implementing the play-types in different orders. This meant that some teachers implemented open-ended play experiences with the children, then a modelled-play experience and finally a purposefully-framed play experience. Other educators commenced with purposefully-framed play, then modelled-play and finished with open-ended play. As we

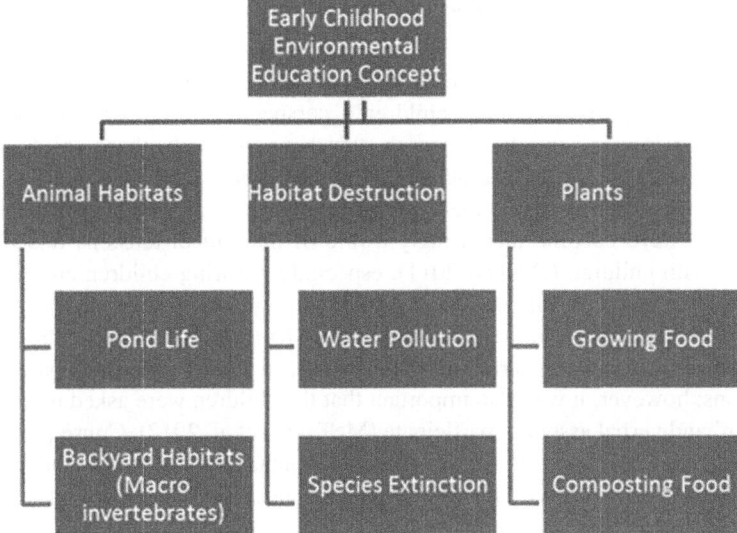

Fig. 1.1 Concept map provided to teachers—what topics might be associated with biodiversity as an early childhood environmental education concept?

discuss in Chaps. 4–6, the order of play-types implemented by Jeanette, Josh and Robyn influenced their thinking about the children's engagement with environmental education. This included how the children were acquiring conceptual knowledge about biodiversity and the provision of opportunities to directly engage children in challenging biophobic (fear of nature) dispositions.

At each setting the teachers progressively implemented a planned play-type experience with the participating children focusing on their selected biodiversity topic. The biodiversity topic was selected by the teachers in collaboration with the children's interests shown at the time. Jeanette selected pond life (Chap. 4), Josh focused on macroinvertebrates (Chap. 5) and Robyn on worms and worm habitat (Chap. 6). The implementation of these experiences were video-taped by the research team and later re-played to the children who were invited to comment on their play and any learning about biodiversity they recalled from their participation. These child video-stimulated recall interviews were also video-taped and the 'second level' video footage was later shown to the teachers during a post play-type implementation interview (Cutter-Mackenzie et al. 2013). During these interviews the teachers reflected on their use of the play-types, including what they believed the children were learning, and how different combinations of play-types appeared to support different aspects of environmental learning. In Chap. 7 we consider how these teacher reflections inform the identification of the two principles for using play-based learning in early childhood environmental education.

1.4 Children as Active Participants

Although the research was specifically about pedagogical strategies implemented by the educators, the participating children's perspectives were critical to understanding the learning involved in each play type. As such, these children were equally involved in the study as active participants who were considered competent to express an opinion about their learning (Lundy et al. 2011). Child focused researchers have become increasingly aware of the ethical tensions raised when working with children (Dockett 2011), especially ensuring children are given the opportunity to authentically decide if they are willing to participate and/or withdraw their assent at any stage of the study (Phelan and Kinsella 2013). All of the children involved in this project participated with the full consent of their parents and guardians; however, it was also important that the children were asked to give their 'written' and verbal assent to participate (McTavish et al. 2012). Consequently, the children were provided with a child-friendly explanatory letter about the research, and invited to read this with their parents/guardians. Children interested in participating indicated their agreement by 'signing' (coloring, making a mark, writing their name) an assent form. At the actual time of each data collection session, the children were verbally re-invited to participate and asked 'would you like to do this today?' and, 'can we make a video of you playing?' Children who said 'no' were able to participate in the activity at a later time if they chose to and were not filmed.

As becomes evident when reading the descriptions of the play-types in Chaps. 4 –6, the children's participation varied even throughout the implementation of the experiences once the filming had commenced. Some of the children started the experiences and then left. Others came in and out of the experiences and were supported to do so by the teachers and the researchers. A child leaving the experience was deemed to have withdrawn 'assent' for that period of time and so was not filmed. In this way, the researchers can be seen to be respecting the children's ethical right to withdraw participation without overt or subtle coercion by adults focused on the collection of data (Sumsion 2003). Instead, this process demonstrates the value placed on the children's meta-narratives as they 'analysed and reported' their own learning witnessed through the video recall of biodiversity focused play experiences (Lundy et al. 2011, p. 716). Children and parents/guardians were invited to either select a pseudonym or indicate a preference for using the child's first name only in the reporting of the research. In all cases except one, children and families elected to use the child's first name.

1.5 Conclusion

The opening vignette for this chapter presents a challenge for early childhood environmental education. This is because swirling some seaweed and playing with some plastic sea animals is simply not enough to help children engage with the

range of knowledge, skills and dispositions needed to support their active participation in society as environmentally engaged citizens. In this book, we draw on the
early childhood education and environmental education literature, in addition to
our own research with teachers and children, to identify two principles for using
play-based learning in early childhood environmental education to address this
challenge. As the examples in Chaps. 4–6 highlight, these principles enable
educators to value the pedagogical potential of different play-types and to combine
the play-types in ways that enable them to realise particular environmental learning
goals for young children. These goals could include engaging with content
knowledge, developing biophilic dispositions towards nature, or learning to value
the environment for its own sake. In the last chapter of the book we use the two
principles to illustrate how Seth's engagement with the children's learning about
biodiversity (opening vignette) could be reconsidered, and in doing so allow him to
realise environmental learning experiences that move beyond noting only the social
and pretend play value associated with the children's play in the wading pool.

References

Cutter-Mackenzie, A., Edwards, S., & Widdop Quinton, H. (2013). Child-framed video research
    methodologies: Issues, possibilities and challenges for researching with children. *Children's
    Geographies.*
Davis, J. (2010). What is early childhood education for sustainability? In J. Davis (Ed.), *Young
    children and the environment: Early education for sustainability* (pp. 21–43). Cambridge:
    Cambridge University Press.
Dockett, S., Einarsdottir, J., & Perry, B. (2011). Balancing methodologies and methods in
    researching with young children (Chap. 5). In D. Harcourt, B. Perry, & T. Waller, (Eds.)
    *Researching young children's perspectives: Debating the ethics and dilemmas of educational
    research with children.* Oxon: Routledge.
Edwards, S., Cutter-Mackenzie, A., & Hunt, E. (2010). Framing play for learning: Professional
    reflections on the role of open-ended play in early childhood education. In L. Brooker &
    S. Edwards (Eds.), *Engaging Play* (pp. 136–151). London: Open University Press.
Edwards, S., & Cutter-Mackenzie, A. (2013). Pedagogical play-types: what do they suggest for
    learning about sustainability in early childhood education? *International Journal of Early
    Childhood, 45*(3), 327–346.
Fleer, M. (2010). *Early learning and development: Cultural-historical concepts in play.*
    Melbourne: Cambridge University Press.
Howitt, C., Lewis, S., & Upson, E. (2011). 'It's a mystery!' A case study of implementing
    forensic science in preschool as scientific inquiry. *Australasian Journal of Early Childhood,
    36*(3), 45–55.
Lundy, L., McEvoy, L., & Byrne, B. (2011). Working with young children as co-researchers: An
    approach informed by the United Nations convention on the rights of the child. *Early
    Education and Development, 22*(5), 714–735.
McTavish, M., Streelasky, J., & Coles, L. (2012). Listening to children's voices: Children as
    participants in research. *International Journal of Early Childhood, 44*, 249–267.
Orr, D. W. (1992). *Ecological literacy: Education and the transition to a postmodern world.*
    Albany: State University of New York.

Phelan, S. K., & Kinsella, E. A. (2013). Picture this…Safety, dignity, and voice—ethical research with children: Practical considerations for the reflexive researcher. *Qualitative Inquiry, 19*(2), 81–90.

Shaffer, L., Hall, E., & Lynch, M. (2009). Toddlers' scientific explorations. Encounters with insects. *Young Children, 64*(6), 18–41.

Sumsion, J. (2003). Researching with children: Lessons in humility, reciprocity and community. *Australian Journal of Early Childhood, 28*(1), 1–9.

Trawick-Smith, J. (2012). Teacher-child play interactions to achieve learning outcomes: Risks and opportunities. In R. C. Pianta, W. S. Barnett, L. M. Justice, & S. M. Sheridan (Eds.), *Handbook of early childhood education*. New York: Giuldford Publications.

UNESCO-UNEP. (1976). The Belgrade Charter: A global framework for environmental education. *Connect: UNESCO-UNEP Environmental Education Newsletter, 1*, 1–2.

Waller, T., Sandseter, E., Wyver, S., Arlemalm-Hagser, E., & Maynard, T. (2010). The dynamics of early childhood spaces: Opportunities for outdoor play? *European Early Childhood Education Research Journal, 18*(4), 437–445.

Wilson, E. O. (1992). *The diversity of life*. Cambridge: Harvard University Press.

Wood, E., & Attfield, J. (2005). *Play, learning and the early childhood curriculum* (2nd ed.). London: Paul Chapman.

Chapter 2
Play-Based Learning in Early Childhood Education

Abstract This chapter problematises play in the twenty-first century and begins with a review of the work of Rousseau, Froebel and Dewey highlighting their enduring influence on play-based practices in early childhood education. The chapter reviews the influence of Piaget's theory on the construction of knowledge via active exploration through play. Working under a Piagetian approach, which has significantly influenced Developmentally Appropriate Practice, the perspective that children learn 'naturally' through play, with the teacher facilitating opportunities for play in the environment, is apparent. However, the authors question whether these views are still current in the twenty-first century, and further question the notion that children learn 'naturally' through play. Applying Vygotsky's understanding about the social mediation of knowledge and learning, and play as a context for adult interaction, the role of the teacher during play to support children's learning is apparent. The authors further question through this reconceptualisation of play: How do teachers know that children are learning? And what is the role of the teacher in children's play? Attention to these questions leads to a more critical consideration of the role of pedagogical play, and the role of the teacher, in early childhood education. This chapter explores such considerations in-depth.

2.1 Introduction

"Recently, I made some profound discoveries that have provided a pivotal moment in my long career within the early childhood profession. The first discovery, which I found in a secret compartment of an old jewellery box, was a yellowed but neatly folded clipping from an early Australian newspaper from the turn of the last century (Fig. 2.1). It detailed the opening of a new kindergarten for the 'small wee ones' in the local Parish Hall. The newspaper spoke in effusive terms of the pencils, paper and dolls provided for the children who were 'seated on wee chairs at little wee tables' under the loving care of the teacher. The second discovery, I realised with a shock, was that nothing much has changed since the

A. Cutter-Mackenzie et al., *Young Children's Play and Environmental Education in Early Childhood Education*, SpringerBriefs in Education, DOI: 10.1007/978-3-319-03740-0_2, © The Author(s) 2014

Fig. 2.1 Australian
newspaper clipping about the
opening of Pingelly
Kindergarten in Western
Australia from the early
1900s (unknown newspaper
and date)

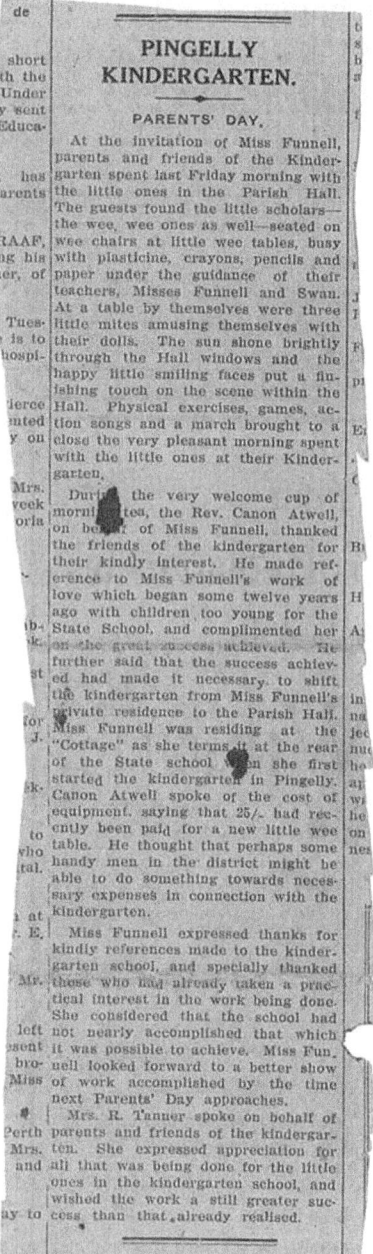

early 1900s in the contemporary early childhood settings and provisions of the twenty-first century. This seems true I realised, despite more than 100 years of intense investigation and research into children's play, learning and education. Why, I wondered with this concrete evidence in my hand, is the early childhood community so resistant to change?" (Moore, personal communication, 20th June, 2012)

A kindergarten established over 100 years ago is filled with small chairs, paper, pencils and dolls. Many years later, this very description remains a familiar account of typical early childhood provision. In this chapter we examine the concept of 'pedagogical play' and how this concept has informed the use of materials such as those celebrated in the opening of the Pingelly Kindergarten in Western Australia in the early 1900s—and continues to inform approaches to learning and teaching in early childhood education today. Pedagogical play refers to the use of play in early childhood education to promote the learning of young children (Wood 2010). Pedagogical play has a long and contentious history in early childhood education, beginning with the argument that children learn most 'naturally' from play, and focussing more recently on problematising what and how children learn through play. In this chapter we consider this history and outline where ideas about the naturalness of children's play came from and how these ideas have more recently been challenged by 'postdevelopmental' perspectives on pedagogical play.

2.2 Historical Theoretical and Philosophical Informants to Early Childhood Education

Deeply entrenched within the historical roots of early childhood education, play has long been a dominant feature of Western-European pedagogy (Rogers 2011). Over many centuries, philosophers, theorists, educationalists and more recently, policy makers have worked hard to define the nature of childhood, play and the purposes of education (Fisher 2008). In particular, researchers have become increasingly interested in how traditional and contemporary theories on play and childhood have informed conceptualisations of childhood (Grieshaber and McArdle 2010), the 'image of the child' (Malaguzzi 1994), and the development of early childhood curriculum (Graue 2008). Wood and Attfield (2005) claim that until the nineteenth century, "childhood was seen as an immature form of adulthood and children from all social classes had little status in society" (p. 29). Wood and Attfield suggest that it was the studies of classical play theorists, such as Rousseau, Froebel and Dewey, that dramatically changed societal views and attitudes towards children, to the extent that "freedom to learn could be combined with appropriate nurturing and guidance" (p. 29), through the strongly held belief that play was critical to children's learning and development (Platz and Arellano 2011).

These early theorists were strong advocates for children learning in, and from, nature as active learners, suggesting that "children learned best when they were allowed to observe and interact with nature and life" (Platz and Arellano 2011, pp. 56–57). Integral to their beliefs, was the view that children were naturally good, and so educational and social goals for young children should be orientated towards nurturing this natural innocence. Platz and Arellano (2011) suggest that "the origins of many early childhood education theories and practices today can be traced back in time to early educators and philosophers who had a passion for the development and education of young children" (p. 54). However, despite the fact that the philosophies of these theorists were not always endorsed during their lifetimes (due to various political and moral stances of the time), their work has clearly impacted on European-Western ideologies regarding the importance of play as a primary mode of learning for young children in early childhood education (Lillemyr 2009).

Jean Jacques Rousseau (1712–1778), as one of the first notable philosophers, was attributed with many idealistic views about children and childhood. Notions associated with child-centred education where "nature requires children to be children first" are believed to have initiated from Rousseau's theories on education (Platz and Arellano 2011, p. 56). Rousseau is known for his romantic views on children's innocence and the 'golden age of childhood' together with other significant shifts in the concept of childhood, as James et al. (1998) suggest:

> Rather than just instilling a sense of childhood innocence, Rousseau, more significantly, opened up the question of the child's particularity, a question that remains central in the status of person, a specific class of being with needs and desires and even rights. And it is this personification which has paved the way for our contemporary concern about children as individuals (p. 13).

While the issue of children's rights appears to have foundations in Rousseau's pioneering work, it has only become a "special safeguard for children" with the *United Nations Convention on the Rights of the Child* (1989) in more recent times (Lee 2001). Rousseau's projection of childhood innocence also paved the way for an image of the innocent child needing protection, and a tendency for adults to feel the need to "shelter children from the corrupt surrounding world ... by constructing a form of environment in which the young child will be offered protection, continuity and security" (Dahlberg et al. 1999, p. 45). Early childhood settings have been perceived as providing this protective role, especially in relation to environmental education which has been viewed as a potentially overwhelming topic for the developmental capacity of young children (see for example Duhn 2012). Graue (2008) argues that this situation is a misplaced function of early childhood education's concern with children's development, and the sense that very young children in their innocence may not be ready to engage with complex conceptual or socially-based ideas.

Oelkers (2002) in his study of *Rousseau and the image of 'modern education'*, claims that Rousseau "took for granted that the self-development of the child is driven by immediate interests, i.e., not by instruction or by formal education" (p. 683), and continues this line of Rousseau's thinking by stating, "If educators let the child always be himself, attending to only what touches him immediately, then and only then will they find the child learning, capable of perceiving, memorizing, and even reasoning" (p. 683). The underlying premise intrinsic to many early childhood philosophies and policies of 'taking the child's interest' clearly has its roots in this theory of Rousseau's approach to early education (Hedges et al. 2011).

It is generally agreed that the theories espoused by the German theorist, Fredrich Froebel (1782–1852) as the creator of the first 'kindergarten' or 'children's garden', were not only the most significant during his time, but still have an enduring influence on current early childhood practices (Ailwood 2007). Sherwood and Reifel (2010) comment on the "central element" of United States kindergartens initially holding "tightly to its Froebelian roots" (p. 323). These roots can likewise be viewed across many Western-European orientated approaches to early childhood curriculum, including in the New Zealand *Te Whariki* (Ministry of Education 1996) early childhood guidelines, the Australian *Early Years Learning Framework* (Department of Education, Employment and Workforce Relations 2009), the Singaporean Curriculum Framework for Kindergartens *Nurturing Early Learners* (Ministry of Education 2012), the framework for the *Early Years Foundation Stage* in the United Kingdom (Department for Education 2012), the *National Curriculum Guidelines on Early Childhood Education and Care in Finland* (Ministry of Social Affairs and Health 2004) and in the American National Association for the Education of Young Children's (NAEYC) *Developmentally Appropriate Practice Guidelines* (2009). In each of these documents reference is commonly made to children's play and their play-based interests as an initial site for learning and development. For example, the Singaporean curriculum document suggests:

> Play is the primary mechanism through which children encounter and explore their immediate environment. As such, play becomes a natural way to motivate children to learn about themselves and the world around them (Ministry of Education 2012, p. 34).

Likewise the Developmentally Appropriate Practice Guidelines say of play:

> Children of all ages love to play, and it gives them opportunities to develop physical competence and enjoyment of the outdoors, understand and make sense of their world, interact with others, express and control emotions, develop their symbolic and problem-solving abilities, and practice emerging skills (NAEYC 2009, p. 14).

Froebel believed that children would learn through their play, and therefore, "learn to live in harmony with others and nature" (Platz and Arellano 2011, p. 60). Edwards and Hammer (2006) suggest that:

> Froebel devised curriculum materials and a methodology of education that would foster a blossoming of concepts and understanding in young children's thinking. His approach to early childhood teaching emphasized the inherent nature of children's learning that unfolds through their play; the structure of developing concepts that were drawn from nature and the role of the teacher. Froebel's understanding of children's play was extrapolated as 'serious work' and he developed a sequence of 'Gifts' and 'Occupations' to harness what he described as a natural energy that could be directed towards learning concepts (p. 195).

The importance Froebel placed on the concepts of "first hand experiences and self-chosen activities" were manifestations of his belief that adults should "begin where the learner is" and only "sensitively intervene" in children's play (Wood and Attfield 2005, p. 29). Many of these ideas are still evident in the philosophies and teaching techniques associated with early childhood education today (Krogh and Slentz 2010). For example, Liebschner (1993) highlights Froebel's theories around the importance of meaningful play embedded in his gifts, occupations and practical 'work in the garden' by quoting Froebel's actual tenet as:

> Play must always be in agreement with the total life of the child as well as with his environment, and cannot stand in isolation or be divorced from it; play will then be educative, serious and meaningful. Through it, life becomes more relevant (p. 54).

Interestingly, Froebel's appeal for play to be in agreement with the child's life could be viewed as a harbinger of the cultural historical argument regarding the significance of context in children's learning. For example, Vygotsky (1997) also talked of the need for educational experience to connect strongly with children's life experiences, saying that "Ultimately only life educates, and the deeper that life, the real world, burrows into the school, the more dynamic and the more robust will be the educational process" (p. 345). Froebel however, was especially interested in implementing his kindergarten ideas and practices for young children in the "space between home and school" as a "half day educational service" (Ailwood 2007, p. 53).

May (2006) argues that since Froebel's times, early childhood advocates have been attempting to "persuade society in general and politicians in particular as to the benefits of early childhood care and education for children prior to school entry" (pp. 245–246). May also suggests early childhood has "always been a site for experiment" (p. 262), and that indeed, to be considered "Froebelian, is about being an advocate for children, for women and for social justice" (p. 262). For Froebel, the image of the child, was one that focused on understanding "the young child as nature" where children's learning and "inherent capabilities" unfolded naturally when given the opportunity to do so (Dahlberg et al. 1999, p. 46).

John Dewey (1859–1952), an American philosopher and educational reformist, believed it was important to provide many different experiences to enable children's learning through play "as a lifelong process in which children grew and learned along the way" (Platz and Arellano 2011, p. 56). Dewey, similar to the philosophers before him, strongly believed in connecting with the natural interests and activities of young children, such that the "question of education is the question of taking hold of his [sic] activities, of giving them direction" (Dewey 1956, p. 36). Interests, play experiences and opportunities for exploring the outdoors arguably placed the child at the centre of education and emphasised learning in social and meaningful contexts (Dewey 1956, p. 33). Wood and Attfield (2005) argue that Dewey viewed children as "co-constructors of their learning; he saw them as active agents and active participants in shaping their learning environments and experiences" (p. 30). Years later these same ideas were to become visible in the Reggio Emilia early childhood education practices, particularly in the focus on the competent and capable child "as an architect of their own learning" (Dodd-Nufrio 2011, p. 236). Interestingly, the concept of the socially agentive child reflects many of the newly emerging ideas from the sociology of childhood (Bass 2010; Corsaro 2011) and the late 1990s positioning of the child as "the child as a co-constructor of knowledge, identity and culture" (Dahlberg et al. 1999, p. 48).

Further changes in societal attitudes towards children and childhood were influenced by increasing childhood studies during the early to mid-twentieth century. Wood (2007) describes the historical trend of increasingly combining theory and practice in early child development and education, providing the prime example of E.R. Boyce (1946) in the "child-centred educational experiment" she set up in London, and states:

> Child-centred education incorporated care, rescue and correction of 'defects' alongside a commitment to free choice and free play within a richly resourced learning environment. There was no distinction between work and play … Content knowledge was embedded in play activities that reflected their everyday lives, and promoted fantasy and imagination (p. 121).

Although, the educational reforms and curriculum created and implemented by the early theorists are relevant to the time and contexts in which they were developed it is clear that much of their early beliefs and images of childhood have had a powerful impact on our current early childhood education systems and practices (Lim and Genishi 2010). In the latter part of the twentieth century political, social and economic changes and pressures became progressively more controlling in how early childhood curriculum was approached, with increasing demands to produce children who would be a "well prepared workforce for the future" (Dahlberg et al. 1999, p. 45). The French historian, Aries in his work on *Centuries of Childhood* (1962) may have "rediscovered the lost childhood from the past" (Frijhoff 2012, p. 24), however the society of the mid-twentieth century also discovered "the child as labour market supply factor" (Dahlberg et al. 1999, p. 46).

Jean Piaget (1896–1980), a Swiss developmental psychologist, was particularly interested in young children's cognitive development. Many aspects of Piaget's theory became associated with early childhood education during the 1960s. This is possibly because of the extent to which his ideas regarding the children's construction of knowledge aligned with existing ideas about the naturalness of children's learning through play already in place due to the influence of Frobel and Rousseau (Krogh and Slentz 2010). Piaget's emphasis on the explorative capacities of young children combined with the suggestion that learning experiences were most appropriately matched with children's play-based stages of development had significant implications for the pedagogical strategies associated with many early childhood programs over the past 50 years (Hatch 2010). Dahlberg et al. (1999) argue that the image of "Piaget's child" as progressing biologically through stages towards maturity was preferred by the scientific and psychological disciplines, suggesting that "the dominant developmental approach to childhood provided by psychology, is based on the idea of natural growth… childhood therefore is a biologically determined stage on the path to full human status" (p. 46).

2.3 Developmentally Appropriate Practice in Early Childhood Education

Piagetian theory and philosophical ideas about children and childhood subsequently informed the influence of the Developmentally Appropriate Practice (DAP) guidelines for early childhood education. Initially published in 1987 by the American National Association for the Education of Young Children (Bredekamp 1987), the guidelines were intended to respond to pressures to make the early childhood curriculum overly academic and provide a theoretical and research evidence base for protecting children's opportunity to learn and develop through the provision of traditionally valued play-based experiences. There was an early emphasis on the provision of play experiences that would support children's active engagement in play and the matching of children's developmental capacities to play activities (Edwards 2003).

Later, significant critique of the DAP guidelines (Kessler 1991; Silin 1987) saw them modified in 1997 (Bredekamp and Copple 1997), and again in 2009, to include greater focus on the role of social and cultural interactions on children's learning, play and development. Nonetheless, many of the Piagetian ideals about early learning and development associated with the DAP guidelines have become firmly entrenched in understandings about appropriate early childhood education. Hatch (2010) attempts to explain why the early childhood field has been hesitant to leave the security of a Piagetian theoretical framework behind:

> It feels heretical to challenge the Piagetian orthodoxy of the early childhood field… [it is] difficult to say why Piaget's core ideas and assumptions of developmental approaches have endured… perhaps early childhood educators have associated the precepts of Piagetian developmentalism so closely with a 'child centred' approach that to abandon them would feel tantamount to abandoning their concern for children (pp. 266–267).

Hatch's view is possibly accurate given the particularly close links established between child-centred practice, the images of the child and the underlying premise for developmental theories. However, emerging pedagogical practices and research interests from outside Piagetian ideas saw increased interest in alternative viewpoints on young children's play and its role in early childhood education. In particular, research began to be directed towards questions such as 'what do young children learn through play?' And 'are children able to learn through free play alone?' (Gibbons 2007; Hedges 2010).

These concerns have recently been summarised by Yelland (2011) who suggested that learning through play can be "problematic and misleading" (p. 5) because whilst children may be having "fun participating in such free play sessions" the type of learning taking place may not necessarily be obvious. The opening vignette to this book in which Seth observed the children swirling seaweed is a case in point. In such situations it is possible to ask "what connections are being made to the child's lived experiences and knowledge building and how are these articulated and extended in supporting activities?" (Yelland 2011, p. 5). The 'problem' with play became highly debated as researchers emphasised the need for adult interaction during children's play to support learning (Winsler and Carlton 2003). Others criticised adult intervention in play as damaging to children's self-agency (O'Brien 2010), and still others worked to promote an understanding of balanced or integrated play that provided opportunities for both child-centred activity and adult interaction (Wood 2013). Meanwhile, the updated Singaporean Curriculum Framework for Kindergartens directly referenced a continuum perspective on children's play, emphasising the role of teacher interactions during play to support children's learning:

> Play can range from being unstructured with free choice by children and no/little active adult support to being highly structured with teacher-led instruction and direction. While recognising the benefits of child-initiated and free play-choice play, this framework highlights the critical role of the teacher in purposeful play (Ministry of Education, Singapore 2012, p. 34)

In part, the problem can be attributed to what was hinted at in the beginning of this chapter—somewhat unchanged materials and practices in the provision of early years pedagogy that mean it can be difficult to change what actually happens in terms of using play as the basis for supporting learning. Krieg (2010) taps neatly into this problem discussing the influence the 'technologies' (i.e., pencils, paper and dolls) of traditional kindergartens have on taken-for-granted pedagogies—that is, the assumption that the provision of stimulating materials will be sufficient for promoting the type of play that will allow children to learn and construct their own understandings of the world. This is the very basis of the play provision offered by Seth in the opening chapter. The plastic sea animals, seaweed and sponges supposedly embed concepts about biodiversity into the play experience—by making these materials available Seth may well believe the children will learn what characterises the different creatures. Meanwhile, recent arguments continue to suggest that, whilst challenging, the "time is ripe for a critical empirical and theoretical look at the contribution of play and an examination of what is perceived as play from the perspectives of all the stakeholders" (Stephen 2010, p. 19). This movement towards a more critical consideration of the role of pedagogical play in early childhood education has commenced within the context of postdevelopmental perspectives on early childhood education.

2.4 Postdevelopmental Perspectives on Early Childhood Education

Continued engagement with ideas associated with Developmentally Appropriate Practice in early childhood education were supported by a range of contemporary perspectives on early learning, development and play, including post-modernism, post-structural, sociocultural and sociology of childhood viewpoints (Nolan and Kilderry 2010). Collectively, these perspectives increasingly captured the notion of being 'postdevelopmental' (Blaise 2009), whilst individually they are understood to hold quite significant theoretical and philosophical lines of thought that distinguish each from the other.

Amongst the most significant of the postdevelopmental perspectives has been the work of the Russian psychologist Lev Vygotsky (1896–1934). Vygotsky developed his theory during the early part of the twentieth Century through periods of great social upheaval and war. Nonetheless, his work has had far reaching implications for early childhood education and contemporary childhood studies in terms of his explanation of children's mastery of play, the development of imagination and the increasingly significant role of the teacher in children's learning (Kozulin 2001; Bodrova 2008). Kozulin et al. (2003) describe Vygotsky's work as:

> At the heart of Vygtosky's theory lies the understanding of human cognition and learning as social and cultural rather than individual phenomena… Vygotsky strongly believed in the close relationship between learning and development and in the sociocultural nature of both. He proposed that a child's development depends on the interaction between a child's individual maturation and a system of symbolic tools and activities that the child appropriates from his or her sociocultural environment. Learning in its systematic, organized, and intentional form appears in sociocultural theory as a driving force of development, as a consequence rather than a premise of learning experiences (p. 1).

Corsaro (2011) supports Vygotsky's ideas about children's interpretation of their culture through the acquisition of language and other cultural "tools or signs" (such as, drawing, objects) which are "created over the course of history and change with cultural development" (p. 15). According to Vygotsky, children "through their acquisition and use of language, come to reproduce a culture that contains knowledge of generations" (Corsaro 2011, p. 15). Corsaro (2011) continues stating "Vygotsky saw practical activities developing from the child's attempts to deal with everyday problems. Furthermore, in dealing with these problems, the child always develops strategies collectively—that is, in interaction with others" (p. 16).

From a sociocultural perspective, the teachers' role is much more proactive and engaged than previous understandings of pedagogical play which tended to highlight the role of the child's freely-chosen investigation in learning. From this perspective, Seth's approach, in which he stood and watched as the children played in the wading pool, would be considered insufficient for supporting learning. The increased role of the adult in children's learning therefore challenged conventional ideas about the child being the 'centre' of learning (Graue 2008), and resulted instead in arguments about pedagogical play that increasingly emphasised adult interactions to support children's conceptual learning and the acquisition of content knowledge (Eun 2010; Fleer 2010). Göncü and Gaskins (2011) argued that this movement represented a feasible reading of Vygotsky's ideas about the social orientation of play, however, noted that the adult "harnessing" of play for educative purposes shifted children's play from a focus on symbolic exploration to an intentional focus on learning (p. 55).

This shift was seen in the uptake of the idea of 'intentional teaching' (Duncan 2009; Epstein 2007) and in the use of the term 'sustained shared thinking' (Siraj-Blatchford 2009), where the educator and child engage in conversation to further promote learning. Questions about how to best balance the role of intentional teaching with children, as opposed to setting up environments for open-ended play and acting as a facilitator to children's learning are becoming increasingly evident in research with educators (Thomas et al. 2011). This is particularly so where educators are concerned with the content associated with young children's learning and how such content learning can be best supported in early childhood contexts. Increasingly it is understood that content knowledge is constructed by children in concert with educators who already hold some degree of knowledge themselves. As Hedges and Cullen (2005) suggest:

> The kinds of informal, everyday knowledge children construct are mediated by teachers' domain knowledge in the context of responsive pedagogical approaches and can be a foundation for the co-construction of more formal knowledge (p. 5).

How children access content through pedagogical play is an area of research that increasingly highlights the relationship between children and teachers as a basis for learning (Hatch 2010). Whilst play and opportunities for freely-chosen play are historically valued and important, content knowledge and how this is co-constructed between children and teachers, is also considered increasingly significant in early childhood education. As Pramling-Samuelsson and Carlsson (2008) argued, if play is to be considered educative in basis it would have to teach children 'something'. This representation of the 'something' sums up the tensions associated with contemporary perspectives on pedagogical play in early childhood education and illustrates the need for principles of play-based learning to inform early childhood environmental education. Otherwise, the situation can be very like the opening vignette in this book in which Seth observed the children at play, but there was little sense of what they learning about biodiversity that was going to contribute to their environmental education.

A shifting emphasis on the nature of interactions between children and adults in early childhood settings suggests instead that content needs to be more explicitly engaged by teachers for the pedagogical potential of play to be realised as environmental learning. Pedagogical play (encompassing the idea that play can be used in early childhood education to support learning) therefore centres on the debate regarding the extent to which the play should be relatively open-ended and exploratory, and the extent to which it should involve focussed interactions between children and adults in relation to particular content (Fleer 2011).

Another area of postdevelopmental research that has contributed to perspectives on pedagogical play is associated with the emergence of ideas from the sociology of childhood perspective (Dahlberg and Moss 2005; Moran-Ellis 2010; Shanahan 2007). James et al. (1998) describe a "new paradigm of the sociology of childhood" where children are no longer merely a "category" but "social actors shaping as well as shaped by their circumstances" (p. 6). James et al. (1998) claim that "the discovery of children as agents" (p. 6) is of prime importance in this new way of thinking about children because it opens opportunities for thinking about how children construct perspectives, experiences and knowledge in relational ways. Dahlberg (2009) also established this perspective, suggesting that knowledge is socially co-constructed by children as social actors capable of creatively influencing their own lives within "their everyday lives in the preschool" (p. 235). Pedagogically, Nolan and Kilderry (2010) argue that:

> Postdevelopmental orientations are inspired by theories and practices located outside child development theory, and suggest that play, and the pedagogical use of play, are not governed by individual children's 'needs'. Instead children are viewed as competent, socially active learners who are able to co-construct their learning intentions, learning strategies and learning outcomes in culturally meaningful ways with peers and adults (p.113).

Similarly, Corsaro (2011) argues that children engaged in peer culture play are able to enact control, autonomy and agency as they negotiate and protect their interactive play spaces within their early childhood settings (p. 161). From a sociology of childhood perspective, educators are likely to view children as competent actors capable of influencing their own learning with ideas and theories of pedagogical worth (Dahlberg et al. 1999, p. 48). With similar beliefs to those expressed by Dewey in the early twentieth century, Dahlberg et al. (1999) defined the 'new' sociology of childhood and the social construction of childhood:

> In this construction of the 'rich' child, learning is not an individual cognitive act undertaken almost in isolation within the head of the child. Learning is a cooperative and communicative activity, in which children construct knowledge, make meaning of the world, together with adults and, equally important, other children: that is why we emphasize that the young child as learner is an active co-constructor. Learning is not the transmission of knowledge taking the child to preordained outcomes, nor is the child a passive receiver and reproducer... he or she is born equipped to learn and does not ask or need adult permission to start learning (p. 50).

Postdevelopmental perspectives on play, whilst emphasising children's co-construction of knowledge in social contexts, also highlight the extent to which play is seen as open to interpretation. This includes seeing play in terms of the impact of gender, peer relationships, cultural experience and socioeconomic opportunities (Grieshaber and McArdle 2010). In early childhood education, this expanded understanding of play has resulted in the suggestion, that rather than seeing pedagogical play only as related to developmental or educational outcomes that educators think about how and why play is being used in early childhood education settings. In this way, play is thought about in terms of the 'context of application' in which it occurs and is used (Brooker and Edwards 2010). This can include developmental and educational outcomes, but also consideration of the impact of peer relationships on children's learning through play or the role of their cultural experiences on learning in early childhood settings. Importantly for early childhood environmental education, the context can and should consider the nature of children's play-based interactions of the world so that these may be orientated towards learning 'something' about the environment.

2.5 Conclusion

Early childhood education has been informed by a rich variety of beliefs and values over many generations of theorists and educators. Many of these ideas are still present in some form in multifaceted combinations of theories, images of the child and pedagogy. Long held views and traditions can be traced from the eighteenth Century through to contemporary thinking about pedagogical play. These include Rousseau's ideas about childhood innocence and protection; Froebel's notion of children being at work when playing in the children's garden; Dewey's focus of the active learner working on real life problems; Boyce's embedding of content knowledge in play; through to Piaget's exposition on the construction of knowledge through active exploration during play. More recently, ideas derived from Vygotsky's understanding about the social mediation of knowledge and learning, and play as a context for adult interaction are increasingly evident in approaches to early childhood education that now also value the role of the educator during play to support learning. The sociology of childhood highlights childhood agency, whilst notions of power relations between children and adults continue to shape discussion regarding the use of play-based learning in early childhood education. While play is gradually reconceptualised, the historical informants are still recognisable, and the Australian kindergarten described in the introduction of this chapter "for the small wee ones", may not be very different from the kindergartens now provided for young children in Singapore, New Zealand, the United States of America, the United Kingdom or Finland. This is not to say that pedagogical practices remain unchanged, rather to reflect on the extent to which early childhood education as a field evolves in relation to highly valued historical ideas about play, and the role of pedagogical play in the education and care of the very young. How these ideas manifest with the provision of early childhood environmental education forms the focus of the Chap. 3.

References

Ailwood, J. (2007). Motherhood, maternalism and early childhood education: Some historical connections. In J. Ailwood (Ed.), *Early childhood in Australia: Historical and comparative contexts*. Frenchs Forest: Pearson Education Australia.

Aries, P. (1962). *Centuries of childhood: A social history of family life*. New York: Knopf.

Bass, L. E. (2010). Childhood in sociology and society: The US perspective. *Current Sociology, 58*(2), 335–350. doi:10.1177/0011392109354248.

Blaise, M. (2009). "What a girl wants, what a girl needs": Responding to sex, gender, and sexuality in the early childhood classroom. *Journal of Research in Childhood Education, 23*(4), 450–460.

Bodrova, E. (2008). Make-believe play versus academic skills: A Vygotskian approach to today's dilemma of early childhood education. *European Early Childhood Education Research Journal, 16*(3), 357–369. doi:10.1080/13502930802291777.

Bredekamp, S. (1987). *Developmentally appropriate practice in early childhood programs serving children from birth through age 8* (Expanded ed.). Washington, D.C.: National Association for the Education of Young Children.

Bredekamp, S., & Copple, C. (1997). *Developmentally appropriate practice in early childhood programs* (Rev ed.). Washington, D.C.: National Association for the Education of Young Children.

Brooker, L., & Edwards, S. (2010). Introduction: From challenging to engaging play. In L. Brooker & S. Edwards (Eds.), *Engaging play* (pp. 1–10). London: Open University Press.

Corsaro, W. A. (2011). *The sociology of childhood* (3rd ed.). Thousand Oaks: Pine Forge Press.

Dahlberg, G. (2009). Policies in early childhood education and care: Potentialities for agency, play and learning. In W. A. Corsaro, M.-S. Honig & J. Qvortrup (Eds.), *The Palgrave handbook of childhood studies*. Basingstoke: Palgrave Macmillan.

Dahlberg, G. & Moss, P. (2005). The ethics and politics in early childhood education. Oxfordshire: Routledge Falmer.

Dahlberg, G., Moss, P., & Pence, A. R. (1999). *Beyond quality in early childhood education and care: Languages of evaluation*. New York: Routledge.

Department of Education. (2012). *Statutory framework for the early years foundation stage: Setting the standards for learning, development and care for children from birth to five*. London: Crown.

Department of Education, Employment and Workplace Relations (DEEWR). (2009). *Belonging, being and becoming: The early years learning framework for Australia*. Canberra: Commonwealth of Australia.

Dewey, J. (1956). *The child and the curriculum, and the school and society*. Chicago: Chicago University Press.

Dodd-Nufrio, A.T. (2011). Reggio Emilia, Maria Montessori, and John Dewey: Dispelling teachers' misconceptions and understanding theoretical foundations. *Early Childhood Education Journal, 39*(4).

Duhn, I. (2012). Making 'place' for ecological sustainability in early childhood education. *Environmental Education Research, 18*(1), 19–29. doi:10.1080/13504622.2011.572162.

Duncan, J. (2009). *Intentional teaching*. Retrieved from http://www.educate.ece.govt.nz/learning/exploringPractice/InfantsandToddlers/EffectivePractices/IntentionalTeaching.aspx

Edwards, S. (2003). New directions: Charting the paths for the role of sociocultural theory in early childhood education and curriculum. *Contemporary Issues in Early Childhood, 4*(3), 251–266.

Edwards, S., & Hammer, M. (2006). The foundations of early childhood education: Historically situated practice. In M. Fleer (Ed.), *Early childhood learning communities: Sociocultural research in practice* (Vol. 8, p. 237). Frenchs Forest: Pearson Education Australia.

Epstein, A. S. (2007). *The intentional teacher: Choosing the best strategies for young children's learning*. Washington, D.C.: National Association for the Education of Young Children.

Eun, B. (2010). From learning to development: A sociocultural approach to instruction. *Cambridge Journal of Education, 40*(4), 401–418. doi:10.1080/0305764X.2010.526593.

Fisher, J. (2008). *Starting from the child: Teaching and learning in the foundation stage* (3rd ed.). Maidenhead: McGraw Hill.

Fleer, M. (2010). *Early learning and development: Cultural-historical concepts in play*. Melbourne: Cambridge University Press.

Fleer, M. (2011). Conceptual Play: foregrounding imagination and cognition during concept formation in early years education, *Contemporary Issues in Early Childhood, 12*(3), 224–240.

Frijhoff, W. (2012). Historian's discovery of childhood. *Paedagogica Historica, 48*(1), 11–29. doi:10.1080/00309230.2011.644568.

Gibbons, A. (2007). The politics of processes and products in education: An early childhood metanarrative crisis? *Educational Philosophy and Theory, 39*(3), 300–311. doi:10.1111/j.1469-5812.2007.00323.x.

Goncu, A., & Gaskins, S. (2011). Comparing and extending Piaget and Vygotksy's understandings of play: Symbolic play as individual, sociocultural, and educational interpretation. In A. Pellegrini (Ed.), *Oxford handbook of the development of play*. New York: Oxford University Press.

Graue, E. (2008). Teaching and learning in a post-DAP world. *Early Education and Development, 19*(3), 441–447. doi:10.1080/10409280802065411.

Grieshaber, S. & McArdle, F. (2010). *The trouble with play*. Maidenhead: Open University Press.

Hatch, J. A. (2010). Rethinking the relationship between learning and development: Teaching for learning in early childhood classrooms. *Educational Forum, 74*(3), 258–268. doi:10.1080/00131725.2010.483911.

Hedges, H. (2010). Whose goals and interest? The interface of children's play and teachers' pedagogical practices. In L. Brooker & S. Edwards (Eds.), *Engaging play* (pp. 25–39). London: Open University Press.

Hedges, H., & Cullen, J. (2005). Subject knowledge in early childhood curriculum and pedagogy: Beliefs and practices. *Contemporary Issues in Early Childhood, 6*(1), 66–79.

Hedges, H., Cullen, J., & Jordan, B. (2011). Early years curriculum: Funds of knowledge as a conceptual framework for children's interests. *Journal of Curriculum Studies, 43*(2), 185–205. doi:10.1080/00220272.2010.511275.

James, A., Jenks, C., & Prout, A. (1998). *Theorizing childhood*. Cambridge: Polity Press in association with Blackwell Publishers Ltd.

Kessler, S. A. (1991). Early childhood education as development: Critique of the metaphor. *Early Education and Development, 2*(2), 137–152.

Kozulin, A. (2001). *Psychological tools: A sociocultural approach to education*. Cambridge: Harvard University Press.

Kozulin, A., Gindis, B., Ageyev, V. S., & Miller, S. (2003). *Vygotsky's educational theory in cultural context*. London: Cambridge University Press.

Krieg, S. (2010). The professional knowledge that counts in Australian contemporary early childhood teacher education. *Contemporary Issues in Early Childhood, 11*(2), 144–155.

Krogh, S., & Slentz, K. (2010). *Early childhood education: Yesterday, today, and tomorrow* (2nd ed.). New York: Routledge.

Lee, N. (2001). *Childhood and society: Growing up in an age of uncertainty*. Philadelphia: Open University Press.

Liebschner, J. (1993). Aims of a good school: The curriculum of Friedrich Froebel: Edited highlights from Froebel's writings. *Early Years: An International Journal of Research and Development, 14*(1), 54–57.

Lillemyr, O. F. (2009). *Taking play seriously: Children and play in early childhood education— an exciting challenge*. Charlotte: IAP, Information Age Pub.

Lim, S., & Genishi, C. (2010). Early childhood curriculum and developmental theory. In P. L. Peterson, E. L. Baker, & B. McGaw (Eds.), *International encyclopedia of education* (3rd ed., pp. 514–519). Oxford: Elsevier.

Malaguzzi, L. (1994). Your image of the child: Where teaching begins. *Early Childhood Educational Exchange, 96*, 52.

May, H. (2006). 'Being Froebelian': An Antipodean analysis of the history of advocacy and early childhood. *Journal of the History of Education Society, 35*(2), 245–262.

Ministry of Education. (1996). New Zealand curriculum document Te Whariki.

Ministry of Social Affairs and Health. (2004). *National curriculum guidelines on ealy childhood education and care in Finland*. Retrieved from http://www.thl.fi/thl-client/pdfs/267671cb-0ec0-4039-b97b-7ac6ce6b9c10

Moran-Ellis, J. (2010). Reflections on the Sociology of Childhood in the UK. *Current Sociology, 58*(2), 186–205. doi:10.1177/0011392109354241.

National Association for the Education of Young Children (NAEYC). (2009). *Developmentally appropriate practice guidelines: Position statement*. Washginton, D.C.: NAEYC.

Nolan, A., & Kilderry, A. (2010). Postdevelopmentalism and professional learning: Implications for understanding the relationship between play and pedagogy. In L. Brooker & S. Edwards (Eds.), *Engaging play* (pp. 108–122). London: Open University Press.

O'Brien, L. (2010). Let the wild rumpus begin! The radical possibilities of play for young children with disabilities. In L. Brooker & S. Edwards (Eds.), *Engaging play* (pp. 182–195). London: Open University Press.

Oelkers, J. (2002). Rousseau and the image of 'modern education'. *Journal of Curriculum Studies, 34*(6), 679–698. doi:10.1080/00220270210141936.

Platz, D., & Arellano, J. (2011). Time tested early childhood theories and practices. *Education, 132*(1), 54–63.

Pramling Samuelsson, I., & Asplund Carlsson, M. (2008). The playing learning child: Towards a pedagogy of early childhood. *Scandinavian Journal of Educational Research, 52*(6), 623–641.

Rogers, S. (Ed.). (2011). *Rethinking play and pedagogy in early childhood education: concepts, contexts and cultures*. Albingdon, England; New York: Routledge.

Shanahan, S. (2007). Lost and found: The sociological ambivalence toward childhood. *Annual Review of Sociology, 33*(1), 407–428.

Sherwood, S. A. S., & Reifel, S. (2010). The multiple meanings of play: Exploring preservice teachers' beliefs about a central element of early childhood education. *Journal of Early Childhood Teacher Education, 31*(4), 322–343. doi:10.1080/10901027.2010.524065.

Silin, J. (1987). The early childhood educator's knowledge base: A reconsideration. In L. G. Katz (Ed.), *Current topics in early childhood education* (Vol. 7, pp. 17–31). Norwood: Ablex Publishing Corp.

Siraj-Blatchford, I. (2009). Conceptualising progression in the pedagogy of play and sustained shared thinking in early childhood education: A Vygotskian perspective. *Educational and Child Psychology, 26*(2), 77–89.

Stephen, C. (2010). Pedagogy: The silent partner in early years learning. *Early Years: Journal of International Research and Development, 30*(1), 15–28. doi:10.1080/09575140903402881.

Thomas, L., Warren, E., & de Vries, E. (2011). Play-based learning and intentional teaching in early childhood contexts. *Australasian Journal of Early Childhood, 36*(4), 69–75.

Vygotsky, L. S. (1997). *Educational psychology*. Boca Raton: St Lucie Press.

Winsler, A., & Carlton, M. P. (2003). Observations of children's task activities and social interactions in relation to teacher perceptions in a child-centered preschool: Are we leaving too much to chance? *Early Education and Development, 14*(2), 155.

Wood, E. (2007). Reconceptualising child-centred education: Contemporary directions in policy, theory and practice in early childhood. *FORUM, 49*(1/2), 119–134.

Wood, E. (2010). Developing integrated pedagogical approaches to play and learning. In P. Broadhead., J, Howard., & E, Wood (Eds.), *Play and learning in the early years*. London: Sage Publications.

Wood, E. (2013). *Play, learning and the early childhood curriculum* (3rd ed.). London: Sage Publications.

Wood, E., & Attfield, J. (2005). *Play, learning and the early childhood curriculum* (2nd ed.). London: Paul Chapman.

Yelland, N. (2011). Reconceptualising play and learning in the lives of young children. *Australasian Journal of Early Childhood, 36*(2), 4–12.

Chapter 3
Environmental Education and Pedagogical Play in Early Childhood Education

Abstract This chapter turns the reader to critical debates and typologies in the environmental education research and literature. Such debates are contextualised within early childhood education and play pedagogies in particular. The authors initially discuss the concepts of sustainable development and sustainability, leading to further critical discussion around the apparent tensions between environmental education and Education for Sustainable Development (ESD)/Education for Sustainability (EFS). The authors challenge the dominant aligning of Education for Sustainability (EFS) and early childhood education, arguing that such alignment is grounded within traditional ideas about children's play. Rather the authors focus upon situating environmental education within contemporary play-based pedagogies. The chapter explores how understanding play-based pedagogy in terms of the role of the teacher is helpful because it widens understandings of 'play' so that content and educator interactions are valued alongside children's activities and interests. Such understandings are essential with respect to supporting children indeveloping ecocentric or biophilic dispositions.

3.1 Introduction

Environmental education is acknowledged as representing a core educational concern in the twenty-first century. This is because environmental education is understood as being an important response to the ways in which human interactions with the world can damage natural and finite resources and put at risk the habitats and ecosystems of different species. In 1972 at the Stockholm *United Nations Conference on the Human Environment*, environmental education was described as "one of the most critical elements of an all-out attack on the world's environmental crisis" (UNESCO-UNEP 1976, p. 2). In the intervening decades, environmental education developed a series of philosophical and research orientated perspectives, in which the purpose of environmental education was variously debated in terms of a range of ideological perspectives (Huckle 1991; Fien 2000;

A. Cutter-Mackenzie et al., *Young Children's Play and Environmental Education in Early Childhood Education*, SpringerBriefs in Education, DOI: 10.1007/978-3-319-03740-0_3, © The Author(s) 2014

Jickling 1992; Jickling and Wals 2007; Sauve 2005). At the international policy level (UNESCO) there has been a notable shift in terminology from Environmental Education to Education for Sustainable Development (ESD) and Education for Sustainability (EFS). Such changes are part of a wider typology of different theoretical and pedagogical positionings or propositions. Sauvé (2005) argues that there are "15 currents" in environmental education whereby sustainable development (including the approaches ESD and EFS) is albeit one current. That argument aside though, the concept of sustainable development (and indeed ESD, EFS among other sustainability education iterations) has unquestionably infiltrated the field of environmental education.

Traditionally 'sustainable development' was defined as "development which meets the needs of the present without compromising the ability of future generations to meet their needs" (World Commission on the Environment and Development 1987, p. 8). However, as with any theory seeking political legitimacy, there are scholars and activists who oppose the ideas underpinning sustainable development (e.g. Jickling and Spork 1998; Selby 2009). One criticism of EDS/EfS is that these approaches derive from an anthropocentric perspective on the environment. An anthropocentric perspective emphasises the use of the environment for human gain, and so sustainability is associated for some scholars with responding to this use so that children become 'agents of change', working to protect the earth's resources from being depleted. Whilst this approach undoubtedly has value (in that children should be supported to understand the importance of not over-using the environment), critics argue that an ecocentric perspective is more appropriate. This is because ecocentrism seeks to value the environment for its own intrinsic value rather than what it offers humans as a resource (Dobson 2007; Eckersley 1992; O'Riordan 1981; Pepper 1984, 1986). Opponents of EfS therefore argue that EfS does not necessarily promote learning to value the environment for its own sake, nor allow children the option of developing their own worldviews about their relationship with the environment (see for example, Kopnina 2012). Hovardas (2013) argues:

> Belief in the intrinsic value of nature, namely, the value nature possesses independently of human valuers, is a strong indication of departing from anthropocentrism (i.e., justification of human conduct only in relation human motives and desires (Curry 2006, cited in text). Granting intrinsic value to nature is related to an ecocentric conceptualisation, according to which natural systems should be considered as bearers of intrinsic value (Gruen 2002, cited in text). Intrinsic valuation of nature and the adoption of ecocentrism might have a substantial effect on images of nature and sense of play (Korfiatis et al. 2009, cited in text). In this regard, environmental education might influence students' worldview to a substantial extent, rather than simply fostering environmental values. Overall, these reservations refer to the formulation of objectives in environmental education and to a potential controversy between endorsing the call for sustainable solutions and, at the same time, respecting learners' autonomy and self-determination (Wals 2010, cited in text) (pp. 1467–1483).

Thus, whilst ESD and EfS are increasingly evident approaches employed in school-based and public education campaigns, it is important for educators and scholars associated with early childhood education to be aware these approaches represent contested arguments in the broader environmental education literature

(Jickling and Wals 2007). This is not to discredit the role of EfS in helping build awareness about the critical importance of sustainability in educational circles, as clearly this been an important platform for getting environmental issues into the curriculum. Rather, the aim here is to alert those involved in early childhood education about how EfS and environmental education are positioned according to the ideological positions they hold about the environment and human relationships with the environment.

Environmental education has had a presence in primary and secondary education for a number of years, and recently emerged in the field of early childhood education in the form of EfS as an official concern (Littledyke and McCrae 2009). The first UNESCO international workshop on environmental education in early childhood was held in 2007, whilst the 2009 Bonn Declaration was amongst the earliest of international documents to recognise the role of early childhood education in environmental education. The 2007 UNESCO workshop resulted in a significant publication titled *'The contribution of early childhood to a sustainable society'* (Pramling Samuelsson and Kaga 2008), aimed at describing how EfS could be understood, used, taught and learned in early childhood settings. Whilst educators and researchers had been working in the area of early childhood environmental education prior to the release of the Pramling and Kaga (2008) document (see for example the significant works of Elliott and Davis 2009), the document served as a touchstone for increased public discussion and awareness regarding the relationship between the education of very young children and the role of sustainability as a core concern of the twenty-first century (Siraj-Blatchford 2009).

3.2 Environmental Education in Early Childhood

Since the publication of *'The contribution of early childhood to a sustainable society'* (Pramling Samuelsson and Kaga 2008) the notion of EfS in early childhood education has gained traction as the most frequently used term associated with environmental education in the early years. However, given debates in the broader environmental education literature about the ideological positions of different approaches to environmental education there is some concern that early childhood education should also be more open to these discussions (Cutter-Mackenzie and Edwards 2013), and so broaden awareness in the field beyond the concept of EfS into consideration of the educational function of environmental education in the first instance. Interestingly, in the history of early childhood sustainability education, it is the ecocentric, rather than anthropocentric perspective that has been most strongly emphasised. This is because the ecocentric perspective seeks to value the earth for its own sake in a way that aligns with historical beliefs in early childhood education about the significance of outdoor and nature-based play as a vehicle for learning about the environment. Pramling Samuelsson and Kaga (2008) argue this very point:

There is a great deal in the history of early childhood education that aligns with education for sustainability e. g. *integrated curriculum approaches* (interdisciplinary), holism, outdoor play and learning, creating a sense of community, social justice etc. We do not have to create entirely 'new' pedagogies in order to 'do' education for sustainability. There is a tradition that could be built upon at the same time as it has to be renewed in terms of thinking about the content and [the need] to work [in] goal directed [ways] in the early years. It is important to raise the question of what the content in Early Childhood Education should be and also what the objectives have to be for fostering children for a life in Sustainability Development. We were also all convinced (from research) that it is not the traditional school subjects and ways of teaching knowledge that has the best effect on children's learning (p. 8).

This has resulted in the situation in which EfS has become somewhat of a default position for environmental education in early childhood education (even though EfS is more likely to orientate towards anthropocentric environmental position whilst early childhood education tends to express ecocentric tendencies towards outdoor play). Consequently, there has been more focus on educating young children about the importance of sustainability in early childhood education (see for example Duhn 2012; Prince 2010), then there has been on understanding how play-based learning connects with environmental education more broadly. Once again, this problem can be seen in the opening vignette for this book in which Seth's play episode largely echoed traditional beliefs about play-based learning in the outdoors, but lacked opportunities for children to engage with environmental learning that would further help them to understand biodiversity, develop biophilic dispositions towards nature and understand the natural habitat of the sea creatures they were incorporating into their play. Environmental education research suggesting that outdoor play *alone* is insufficient for helping children develop later pro-environmental dispositions as adults underscores the significance of this point (Blanchard and Buchanan 2011).

The need for more focused learning about the environment than that enabled by children's exploratory and outdoor based play is illustrated by the Vadala et al. (2007) study regarding the role of children's outdoor play experiences on their later adult-orientated environmental interests. They conducted extensive interviews with 61 participants aged 18–35 years, some who were involved in professional conservation related employment or volunteer activities. Participants were asked to recall and describe their childhood experiences in the outdoors. Interestingly, Vadala et al. (2007) identified two types of outdoor play, including 'child-nature play' and 'child–child play in nature'. Their findings suggested that children who participated in 'child–child play in nature' were more likely to use things found in nature (such as stones, sticks or walnuts) to play war games or build forts than were children who participated in 'child-nature play'. 'Child-nature play' was characterised by children's interests in collecting frogs, searching under logs for bugs and beetles or capturing fireflies. These adults also reported having their interest in nature actively supported by parents who provided access to books, field guides and magazines on natural history. One participant reported "you would just sit back and read them [field guides] like novels" (Vadala et al. 2007, p. 7). 'Child-nature play' adults were more likely to be involved in professional or volunteer conservation roles than those adults who participated predominately in 'child–child play in nature'. This meant that simply being outdoors was not necessarily enough to foster environmental knowledge or understanding in ways that contributed to meaningful environmental interests and behaviors in later adulthood. What mattered was the child's orientation to nature and the fostering of their interest via content supplied by parents. Being outside was not necessarily equated with understanding nature as for some adults the environment simply provided the resources for their childhood imaginative games and activities.

3.3 Biophilia and Biophobia

The Vadala et al. (2007) findings can be understood in relation to two important concepts in environmental education known as biophilia and biophobia. Biophilia is considered to be children's love of and affinity with nature (Wilson 1992). According to Hyun (2005) "biophilia is a theoretical notion that there is a fundamental, genetically based human need and propensity to affiliate with nature and life" (p. 200). Orr (1992) argues if biophilia is not encouraged and nurtured in the early years of life, the opposite occurs and children can develop a fear of nature which is described as biophobia. In the Vadala et al. (2007) study, opportunities for developing a biophilic disposition may be have been most likely to emerge from the experiences of those children participating in 'child-nature play' because this play was orientated towards meaningful engagement with and learning about nature, rather than simply using what nature offered as a resource for play. Research by Hyun (2005) regarding the ways in which children and adults perceive nature would concur with this suggestion. He found that children tend to engage more directly with nature "by doing more touching, smelling, drawing and pretending in a direct and descriptive manner than adults, who did not actively participate" (p. 205). Thus, a disposition towards biophilia is likely to require active opportunities to engage with nature, supported by later opportunities to engage with information about the experience. Seth's wading pool optimistically filled with sponges, sea weed and plastic sea creatures may in fact work to promote biophobia amongst the children—unless some means for later engagement with content knowledge about these creatures is provided.

An important point about biophilia and biophobia in early childhood education is the extent to which educators themselves are likely to express each disposition, and the consequent impact these dispositions have on educator capacities for engaging children in environmental educational experiences. Figure 3.1 presents two contrasting discussions between a child and educator exhibiting either a biophilic or biophobic attitude. Here, it can be seen that the educator leaning towards biophilia is able to support the child's learning needs with respect to understanding the importance of biodiversity and associated concepts such as habitat.

In these examples, the first educator exhibits a biophilic disposition. Her inclination towards respecting the 'snake' extends to helping the child learn the

Biophilia

Child: I saw a snake in my backyard yesterday

Educator: Aren't they so beautiful how they move?

Child: My Dad said I was very lucky to see a large python. So we took a photo. I asked Dad if I could keep him. I said he could sleep in my room. Dad said I couldn't because his home is in the bush

Educator: I know a book called 'The salamander room' (Mazer, 1991) that is about a little boy who tries to keep a salamander but found he couldn't unless he turned his house into a forest

Child: Can we read that now?

Educator: Sure. Let's tell all the other children about the snake you saw yesterday. I am sure we could find out lots more information on pythons too. About what they eat, where they sleep and so on

Biophobia

Child: I saw a snake in my backyard yesterday

Educator: Did your parents kill it?

Child: No, we took a photo of it

Educator: Did you tell your neighbors? You know snakes are very dangerous. They are poisonous and they bite. They could kill you.

Child: Dad said they are beautiful

Educator: Yeah, beautiful when they are dead

Fig. 3.1 Educator dispositions towards biophilia and biophobia

correct terminology (python) and to offering access to more information about the likely habitat and life needs of the python. In this way, the adult's biophilic disposition increases the likelihood of the child accessing the range of content material that the children in the Valdala et al. (2007) study were provided with by their parents—leading to an experience of nature that built and supported a respect for the environment that carried into adulthood. In contrast, the second educator promotes a view of nature that sees the snake as frightening and dangerous. The opportunity to learn more about the reptile is shut down by the suggestion that such a creature could only be beautiful when it was 'dead'. These examples show how environmental education in early childhood education requires more than providing children with outdoor play experiences in nature. Rather, opportunities for play that involve conversations with adults holding biophilic dispositions can be a necessary precursor to accessing content knowledge. In Chaps. 4, 5 and 6 of this book the biophilic dispositions held by Jeanette, Josh and Robyn were a significant influence on their decision making regarding the provision of content knowledge to the children during modelled and purposefully-framed play.

3.4 Pedagogical Play in Early Childhood Education

In Chap. 2 we outlined how theoretical and philosophical ideas about play have influenced understandings about pedagogy in early childhood education. An important idea in Western-European pedagogy has been that children's learning and development is most effectively supported through participation in open-ended and freely chosen play. This idea connects very strongly with ideas proposed by Piaget regarding children's active construction of knowledge and Froebel's and Dewey's arguments regarding the role of play in the child's life as a vehicle for purposeful learning (Wood and Attfield 2005). These ideas about play are strongly entrenched in understandings about early childhood education that are still typically expressed in curriculum documentation or different 'approaches' to early childhood education. For example, the Developmentally Appropriate Practice guidelines (Copple and Bredekamp 2009) suggest:

> Children of all ages love to play, and it gives them opportunities to develop physical competence and enjoyment of the outdoors, understand and make sense of their world, interact with others, express and control emotions, develop their symbolic and problem-solving abilities, and practice emerging skills (p. 14).

This orientation towards play in early childhood pedagogy continues to resonate with the field, and whilst the presence of play itself in early childhood education has not necessarily been critiqued, how play is used and understood in relation to young children's learning has attracted significant research attention. A particularly important body of play-based literature is focused on what young children are likely to learn whilst playing in early childhood settings. An initial concern in this literature was the extent to which young children were likely to learn content knowledge by participating in open-ended and interest-driven play.

Wood (2007) went to the heart of this concern by questioning the extent to which play could be argued to have an educational function if it relied predominately on children's interests in a way that did not deliberately connect with conceptual knowledge and the content associated with a particular learning area:

> It is not clear whether children's interests are themselves goals, whether children create their own goals through their interests and, if so, what those goals are. A further question focuses on whether educators recognise and act on those interests as personal and/or social goals. For example, whilst playing with materials in a water tray may enable children to observe that objects behave in different ways, they will not spontaneously learn the concept of floating and sinking, volume and mass without educative encounters with more knowledgeable others. In other words, play activities may stimulate learning-relevant processes, but may be content free which juxtaposes the developmental against the educational rationale for play (p. 125).

The line of argument expressed by Wood (2007) was largely initiated against a background of theoretical and philosophical change in early childhood education. Other researchers were raising similar questions and concerns regarding the assumed relationship between children's participation in interest-driven and open-ended play and the learning of content knowledge (Hedges and Cullen 2005;

Kallery and Psillos 2001). These investigations were characterized by interest in ideas derived from the sociology of childhood and sociocultural theory. Now broadly encapsulated in the idea of being 'post-developmental' these ideas were focused on addressing perceived limitations associated with traditional ideas about play-based learning such as those emerging from the works of Piaget, Froebel and Dewey amongst others (see Chap. 2). A core concern was focused on under-standing the child in 'context', rather than focusing on the individual child and the construction of knowledge through play. Context included consideration of the role of relationships in children's learning and increasingly referenced the ways in which social and cultural experiences mediated what and how young children learned. Research investigating children's content learning during play drew on sociocultural ideas about learning and development derived from the work of Vygotsky (2004) and Rogoff (2003). These ideas included an emphasis on the role of the adult during play as a support to children's learning and the importance of children's intent participation during social and cultural activities in learning.

A stream of research emerged focussing on understanding the relationship between children's play and their learning of content during such play in early childhood settings (i.e. Pramling Samuelsson and Asplund Carlsson 2008; Robbins 2003). This research increasingly emphasised the importance of adult interactions during play as a means of supporting children's developing conceptual under-standings as basis for building content knowledge (Jordan 2009). This included the concept of Sustained Shared Thinking (Siraj-Blatchford 2009) which emerged from the Effective Provision of Pre-School Education research conducted in the United Kingdom (Siraj-Blatchford et al. 2008). Sustained Shared Thinking was linked to the provision of high quality early learning experiences for young children and arguably characterised by interactions between children and adults that were focused on building knowledge and ideas in the context of play-based experiences. In Australia, Fleer (2010) proposed the idea of contextual inter-subjectivity during children's play. She suggested that interest-driven and open-ended play was an important and appropriate aspect of early childhood education. However, she argued that educators needed to ensure that they understood the context of chil-dren's play so that they were able to engage and interact with children in ways that supported learning rather than assuming that children were learning particular concepts through the provision of play experiences alone. In the United States the concept of intentional teaching was used to describe the importance of achieving a balance between child and teacher initiated activity and interactions:

> An effective early childhood program combines *both* child-guided and adult-guided educational experiences. The terms 'child-guided experience' and 'adult-guided experi-ence' do not refer to extremes (that is, they are not highly child-controlled or adult-controlled). Rather, adults play intentional roles in child-guided experience; and children have significant, active roles in adult-guided experience. Each takes advantage of planned or spontaneous, unexpected learning opportunities (Epstein 2007, p. 3).

An important aspect of intentional teaching was the inclusion of content knowledge in the interactions children and teachers would have together. In Sweden,

Pramling Samuelsson and Carlsson (2008) noted that children should learn 'something' from their interactions during play. Like Bodrova and Leong (2011), they highlighted how learning 'something' was important for extending children's play so that children would have more knowledge to draw on to inform their play-scripts. Understandings about the relationship between adults, children and content during play-based learning have grown through the use of concepts such as intentional teaching, inter-subjectivity and sustained shared thinking. These concepts have supported the emergence of pedagogical ideas about play-based learning that focus on understanding play across a continuum of activity. In these arguments play is not focused on so much as an interest-driven and freely chosen activity in early childhood education as it is understood pedagogically as an experience encompassing a range of activities, including those that might be solely child-initiated and open-ended to those that are more adult directed and/or initiated. This also includes activities in between either end of the continuum that are likely to include a balance of child to child and adult to child interactions and engagements around both play and content learning.

The continuum idea is expressed in descriptions such as integrated pedagogies (Wood 2013) and pedagogical activity (Dockett 2011) that emphasise the importance of play for children's learning but also acknowledge the extent to which educators are able to support this learning when engaging in meaningful interactions with young children. This orientation towards play is evident in contemporary early childhood curriculum frameworks that refer to the role of the educator in engaging young children's learning. For example, the United Kingdom's Early Years Foundation Stage (Department for Education 2012) suggests:

> Each area of learning and development must be implemented through planned, purposeful play and through a mix of adult-led and child-initiated activity. Play is essential for children's development, building their confidence as they learn to explore, to think about problems, and relate to others. Children learn by leading their own play, and by taking part in play which is guided by adults. There is an ongoing judgement to be made by practitioners about the balance between activities led by children, and activities led or guided by adults. Practitioners must respond to each child's emerging needs and interests, guiding their development through warm, positive interaction (p. 5).

In the Australian Early Years Learning Framework (Department of Education and Employment and Workforce Relations 2009), the balance between adult and child-initiated play as a basis for learning is described as such:

> Early childhood educators take on many roles in play with children and use a range of strategies to support learning. They engage in sustained shared conversations with children to extend their thinking (Siraj-Blatchford and Sylva 2004, cited in text). They provide a balance between child led, child initiated and educator supported learning. They create learning environments that encourage children to explore, solve problems, create and construct (p. 5).

Interest-driven and open-ended play in early childhood education is still highly valued for the social, emotional, cognitive and language benefits it arguably provides for young children. However, as recent research suggests, and curriculum frameworks such as the Early Years Foundation Stage and Early Years Learning

Framework increasingly describe, interest-driven and open-ended play is also complemented by educator initiated experiences and interactions aimed at building the content knowledge associated with children's interests and activities. Trawick-Smith (2012) describes the movement towards intentional teaching in terms of three main approaches to pedagogical play, including the "trust in play approach", the "facilitate play approach" and the "enhance learning outcomes through play approach" (pp. 260–262). The "trust in play approach" involves educators providing children with opportunities to engage in open-ended activity in which content is associated with the nature of the materials provided. The "facilitate play approach" involves educators interacting with children during play to add complexity to play scenarios and to help children identify play content. The "enhance learning outcomes through play approach" involves teachers purposefully identifying content they intend for children to interact with during play in order to meet pre-determined learning outcomes. Trawick-Smith (2012) argues that play is used most effectively when teachers combine the approaches in various ways according to what they learn about children's learning through observation and assessment.

Earlier in this chapter we noted that early childhood environmental education needed to be based on more than children's experiences of outdoor play in nature. This was because research shows that play alone does not help children to develop pro-environmental dispositions and understandings (Davis 2010), and further, that adults disposed toward biophilic attitudes towards the environment help children access the content knowledge that extends nature play into understanding about the environment. The recent emergence in the field of early childhood education of the complementary use of different types of play, including both child and adult initiated play, provides a strong basis for beginning to understand how children's outdoor play may be connected with learning opportunities via educators who are interested in promoting environmental education with young children. This is because contemporary orientations towards play-based learning focus on the inclusion of content during play and the ways in which this play can be engaged by children and adults to support conceptual and content based learning. This means there is potential for considering how different forms of pedagogical play can be used by teachers in early childhood environmental education. As we noted in Chap. 1, the pedagogical play-types we have drawn on to inform our research with teachers include open-ended play, modelled play and purposefully-framed play.

3.5 Conclusion

Environmental education is recognised as a core educational concern for the twenty-first century. In recent years, this recognition has been extended to the field of early childhood education. Environmental education and early childhood education can involve more than aligning the values of EfS with the traditional ideas about children's play. It can also be focused on determining how environmental education can be located in early childhood education in a way that addresses needing to learn 'something' (Pramling Samuelsson and Carlsson 2008) about the environment using play-based learning. Recent advances in understanding play-based pedagogy in terms of intentional teaching are helpful because they widen understandings of 'play' so that both content and educator interactions are valued alongside children's activities and interests. This means there is space for considering environmental education in terms of content, but also in terms of the educator interactions that are necessary for realising this content so that children are supported in the development of pro-environmental dispositions and understandings. In the next three chapters we now consider how Jeanette, Josh and Robyn approached the use of play-based learning in early childhood environmental education.

References

Blanchard, P. B., & Buchanan, T. K. (2011). Environmental stewardship in early childhood. *Childhood Education, 87*(4), 232–238.

Bodrova, E., & Leong, D. (2011). Revisiting Vygotskian perspectives on play and pedagogy. In S. Rogers (Ed.), *Rethinking play and pedagogy in early childhood education: Concepts, contexts and cultures* (pp. 60–73). London: Routledge.

Copple, C., Bredekamp, S., & National Association for the Education of Young Children. (2009). *Developmentally appropriate practice in early childhood programs serving children from birth through age 8* (3rd ed.). Washington, DC: National Association for the Education of Young Children.

Cutter-Mackenzie, A., & Edwards, S. (2013). The next 20 years: Imagining and re-imagining sustainability, environment and education in early childhood education. In S. Elliott, Davis, J., Edwards, S., & Cutter-Mackenzie, A. (Eds.), *Best of Sustainability: Research, Practice and Theory* (pp. 61–67). Deakin West: Early Childhood Australia.

Cutter-Mackenzie, A., Edwards, S., & Widdop Quinton, H. (2013). Child-framed video research methodologies: Issues, possibilities and challenges for researching with children. *Children's Geographies*.

Davis, J. (2010). What is early childhood education for sustainability? In J. Davis (Ed.), *Young children and the environment: Early education for sustainability* (pp. 21–43). Cambridge: Cambridge University Press.

Department of Education, Employment and Workplace Relations, (DEEWR). (2009). *Belonging, being and becoming: The early years learning framework for Australia*. Canberra: Commonwealth of Australia.

Department of Education. (2012). *Statutory framework for the early years foundation stage. Setting the standards for learning, development and care for children from birth to five*. London: Crown.

Dobson, A. (2007). *Green Political thought* (4th ed.). London: Routledge.

Dockett, S. (2011). The challenge of play for early childhood education. In S. Rogers (Ed.), *Rethinking play and pedagogy in early childhood education: Concepts, contexts and cultures* (pp. 32–48). London: Routledge.

Duhn, I. (2012). Making 'place' for ecological sustainability in early childhood education. *Environmental Education Research, 18*(1), 19–29. doi:10.1080/13504622.2011.572162.

Eckersley, R. (1992). *Environmentalism and Political Theory: Toward an Ecocentric Approach*. London: UCL Press.

Elliott, S., & Davis, J. (2009). Exploring the resistance: An Australian perspective on educating for sustainability in early childhood. *International Journal of Early Childhood, 41*(2), 65–77.

Epstein, A. S. (2007). *The intentional teacher: Choosing the best strategies for young children's learning*. Washington, D.C.: National Association for the Education of Young Children.

Fien, J. (2000). Education for the environment: A critique- an analysis. *Environmental Education Research, 6*(2), 179.

Fleer, M. (2010). *Early learning and development: Cultural-historical concepts in play*. Melbourne: Cambridge University Press.

Hedges, H., & Cullen, J. (2005). Subject knowledge in early childhood curriculum and pedagogy: beliefs and practices. *Contemporary Issues in Early Childhood, 6*(1), 66–79.

Hovardas, T. (2013). A critical reading of ecocentrism and Its meta-scientific use of ecology: Instrumental versus emancipatory approaches in environmental education and ecology education. *Science & Education, 22*(6),1467–1483.

Huckle, J. (1991). Education for sustainability: Assessing pathways to the future. *Australian Journal of Environmental Education, 7*, 43–62.

Hyun, E. (2005). How is young children's intellectual culture of perceiving nature different from adults'? *Environmental Education Research, 11*(2), 199–214.

Jickling, B. (1992). Why I don't want my children to be educated for sustainable development. *Journal of Environmental Education, 23*(4), 5–8.

Jickling, B., & Spork, H. (1998). Education for the Environment: A Critique. *Environmental Education Research, 4*(3), 309–327.

Jickling, B., & Wals, A. (2007). Globalization and environmental education: looking beyond sustainable development. *Journal of Curriculum Studies, 40*(1), 1–21.

Jordan, B. (2009). Scaffolding learning and co-constructing understandings. In A. Anning, J. Cullen & M. Fleer (Eds.), *Early childhood education. Society and culture* (2nd ed., pp. 39–53). Los Angeles, CA: SAGE Publications.

Kallery, M., & Psillos, D. (2001). Pre-school teachers' content knowledge in science: Their understanding of elementary science concepts and of issues raised by children's questions. *International Journal of Early Years Education, 9*(3), 165–179. doi:10.1080/09669760120086929.

Kopnina, H. (2012). Education for sustainable development (ESD): the turnaway from 'environment' in environmental education? *Environmental Education Research, 18*(5), 699–717.

Littledyke, M., & McCrea, N. (2009). Starting sustainability early: Young children exploring people and places. In N. Taylor & C. Eames (Eds.), *Education for sustainability in the primary curriculum: a guide for teachers* (pp. 39–57). South Yarra, Vic.: Palgrave Macmillan.

O'Riordan, T. (1981). *Environmentalism*. London: Pion Limited.

Orr, D. W. (1992). *Ecological Literacy: Education and the Transition to a Postmodern World*. Albany: State University of New York.

Pepper, D. (1984). *The roots of modern environmentalism*. London: Groom Helm.

Pepper, D. (1986). *The roots of modern Environmentalism*. London: Routledge.

Pramling Samuelsson, I., & Asplund Carlsson, M. (2008). The playing learning child: Towards a pedagogy of early childhood. *Scandinavian Journal of Educational Research, 52*(6), 623–641.

Pramling Samuelsson, I., & Kaga, Y. (2008). *The contribution of early childhood education to a sustainable society*. Paris: UNESCO.

Prince, C. (2010). Sowing the seeds: Education for sustainability within the early years curriculum. *European Early Childhood Education Research Journal, 18*(3), 273–284. doi:10.1080/1350293X.2010.500082.

Robbins, J. (2003). The more he looked inside the more Piglet wasn't there: What adopting a sociocultural perspective can help us see. *Australian Journal of Early Childhood, 28*(2), 1–7.

Rogoff, B. (2003). *The cultural nature of human development*. Oxford: Oxford University Press.

Sauve, L. (2005). Currents in environmental education: Mapping a complex and evolving pedagogical field. *Canadian Journal of Environmental Education, 10*, Spring, 11–37.

Selby, D. (2009). The Firm and Shaky Ground of Education for Sustainable Development. In B. Chalkley, M. Haigh, & D. Higgitt (Eds.), *Education for sustainable development: Paper in honour of the United Nations decade of education for sustainable development (2005–2014)* (pp. 199–213). New York: Routledge.

Siraj-Blatchford, I., Taggart, B., Sylva, K., Sammons, P., & Melhuish, E. (2008). Towards the transformation of practice in early childhood education: the effective provision of pre-school education (EPPE) project. *Cambridge Journal of Education, 38*(1), 23–36. doi:10.1080/03057640801889956.

Siraj-Blatchford, I. (2009a). Conceptualising progression in the pedagogy of play and sustained shared thinking in early childhood education: A Vygotskian perspective. *Educational and Child Psychology, 26*(2), 77–89.

Siraj-Blatchford, J. (2009b). Editorial: Education for sustainable development in early childhood. *International Journal of Early Childhood, 41*(2), 9–22.

Trawick-Smith, J. (2012). Teacher-child play interactions to achieve learning outcomes – Risks and opportunities. In R. C. Pianta, W. S. Barnett, L. M. Justice, & S. M. Sheridan (Eds.), *Handbook of early childhood education.* USA: Giuldford Publications.

UNESCO-UNEP. (1976). The Belgrade Charter: A global framework for environmental education. *Connect: UNESCO-UNEP Environmental Education Newsletter, 1,* 1–2.

Vadala, C. E., Bixler, R. D., & James, J. J. (2007). Childhood play and environmental interests: Panacea or snake oil? *Journal of Environmental Education, 39*(1), 3–18.

Vygotsky, L. S. (2004). Imagination and creativity in childhood. *Journal of Russian and East European Psychology, 42*(1), 7–97.

Wilson, E. O. (1992). *The diversity of life.* Cambridge, MA: Harvard University Press.

Wood, E. (2007). Reconceptualising child-centred education: Contemporary directions in policy, theory and practice in early childhood. *Forum, 49*(1/2), 119–134.

Wood, E., & Attfield, J. (2005). *Play, learning and the early childhood curriculum* (2nd ed.). London: Paul Chapman.

Wood, E. (2013). *Play, learning & the early childhood Curriculum* (3rd ed.). London: Sage Publications.

World Commission on Environment and Development. (1987). *Our common future.* Oxford: Oxford University Press.

Chapter 4
Jeanette: Pond Life

Abstract This chapter presents Jeanette's (an early childhood teacher) and the children's experiences in implementing the three different play types at Cornish College Early Learning Centre, Melbourne, Australia. Using Jeanette's knowledge of the children's past interests she planned an adventure to the pond (also referred to as the lake) in the grounds of the College as a learning opportunity to teach environmental education. Jeanette chose to focus on investigating concepts of sustainability, biodiversity and animal habitats. She used the play-types in the order of open-ended play, modelled-play to raise questions with the children to stimulate their learning, and purposefully framed play to engage the children with content to build their understanding about biodiversity. The order of play-types suited Jeanette as it was consistent with her typical approach to teaching. Whilst the open-ended play experiences helped Jeanette ascertain the children's existing knowledge base, for her it seemed to misconstrue what the children believed they would find in the pond (for example sharks and crocodiles). The later engagement of collecting the water and finding the creatures in the water in modelled-play and purposefully-framed play led the children to an understanding of the range of creatures that actually lived in the habitat. As such, purposefully framed play created a context for supporting children's understanding of life and supported the development of their own biophilia dispositions alongside Jeanette's disposition.

4.1 Jeanette and the Cornish College Early Learning Centre

Jeanette is an early childhood educator working at Cornish College Early Learning Centre. Cornish College Early Learning Centre is an early childhood education centre attached to a private Uniting Church school for children in kindergarten (aged 3–5 year) up to year ten (16 years) in the outer suburbs of Melbourne, Australia. The School opened in 1984 and from its onset had a strong environmental education and sustainability ethos to which Jeanette was strongly committed:

A. Cutter-Mackenzie et al., *Young Children's Play and Environmental Education*
in Early Childhood Education, SpringerBriefs in Education,
DOI: 10.1007/978-3-319-03740-0_4, © The Author(s) 2014

> The vision of the Cornish campus founder, Richard B. Cornish, was to create a place of education where young people would be given the opportunity to understand and value the wisdom of living and working more sustainably (Cornish College 2012a).

Commencing in kindergarten, the College's educational philosophy is based on the following principles:

- A vision for the whole community of sustainable living based around sustainable thinking dispositions, including personal, socio-cultural, urban/technological and natural dimensions.
- Emphasis on creativity and the development of thinking skills.
- Differentiated curriculum to cater for different learning styles.
- Strong emphasis on building foundation skills for learning through structured inquiry.
- Children and staff work together collaboratively in a team structure (Cornish College 2012a).

In the Early Learning Centre, the College draws explicitly upon the educational principles associated with Reggio Emilia (Edwards et al. 1998) as informants to pedagogy, including:

1. *The child as a protagonist.* Children are rich, strong, and capable. All children have preparedness, potential, curiosity and interest in constructing their learning, negotiating with everything their environment brings to them. Children, teachers and parents are considered the three central protagonists in the educational process.
2. *The child as a collaborator.* Education has to focus on each child in relation to other children, the family, the teachers and the community, rather than on each child in isolation. There is an emphasis on work in small groups.
3. *The child as a communicator.* This approach fosters children's intellectual development through a systematic focus on symbolic representation, including words, movement, drawing, painting, building, sculpture, shadow play, collage, dramatic play and music which leads children to surprising levels of communication, symbolic skills and creativity. Children have the right to use many materials in order to discover and communicate what they know, understand, wonder about, question, feel and imagine. In this way, they make their thinking visible through their many natural languages.
4. *The environment as a third teacher.* The use of space encourages encounters, communication and relationships. Every corner of every space has an identity and a purpose, is rich in potential to engage and communicate and is valued and cared for by the children and the adults.
5. *The teacher as a partner, nurturer and guide.* Teachers facilitate children's exploration of themes, work on short and long term projects and guide experiences of joint, open ended discovery and problem solving. To know how to plan and proceed with their work, teachers listen and observe children closely.

Fig. 4.1 Cornish College grounds

Teachers ask questions; discover children's ideas, hypotheses and theories, and provide occasions for discovery and learning.

6. *The teacher as a researcher.* The teachers see themselves as researchers preparing the documentation of their work with children who they also see as researchers.

7. *The documentation as communication.* Careful consideration and attention are given to the presentation of the thinking of the children and the adults who work with them.

8. *The parent as partner.* The ideas and exchange of ideas between parents and teachers favour the development of a new way of educating, which helps teachers to view the participation of families not as a threat, but as an intrinsic element of collegiality and as the integration of different wisdom (Cornish College 2012b, p. 5; see also Cadwell 1997, pp. 5–6 from which the Cornish College principles were adapted).

The Early Learning Centre and School is situated on 42 hectares of marshland and has a farm, lake, orchard, an island and significant bush land (Fig. 4.1). This outdoor environment is important to learning at the centre and is not dissimilar to a Scandinavian nature kindergarten where Froebel's ideas about play are expressed in values that promote extensive nature and outdoor experiences, and few commercial toys (Waller et al. 2010).

At Cornish College Early Learning Centre information provided for parents includes a description of the outdoor experiences in which children attending the centre will participate:

> To increase their connection to nature, as well as many other educational benefits, the children will be spending extended periods of time outside. This will give them tangible ways of working with nature in a variety of settings (Cornish College 2012b, p. 17).

Fig. 4.2 A copy of
Jeanette's teaching concept
map

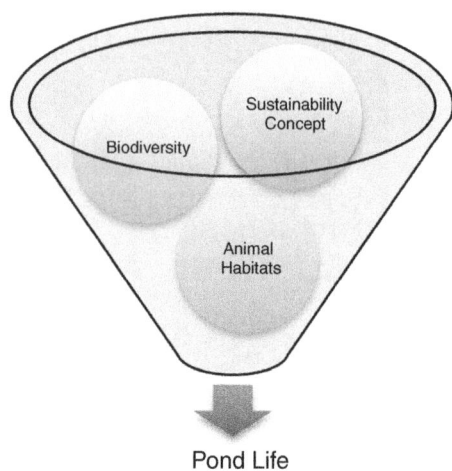

Pond Life

4.2 Focusing on Pond Life

During term two Jeanette noticed that the children were interested in mini-beasts
(macro-invertebrates) whilst having spent some time planting seeds in the vege-
table garden. Jeanette wrote in her teacher journal:

> Interest arose and a group of 5 children formed a project group called 'The Creatures
> House' to explore where some of these creatures live and what they need to stay alive.
> I decided to extend and further the interest in creature and their habitats (Fig. 4.2).

For her participation in the project Jeanette decided to capitalize on the Centre's
strong commitment to children's outdoor experiences which meant she had access
to a lake located on campus within walking distance of her classroom. This lake
was quite a large body of water, and so considered larger than a pond. The term
'pond life' however is used to describe all aquatic creatures living in the lake.
Jeanette believed that the lake would provide an appropriate opportunity for the
children to engage in learning more about macro-invertebrates in way that would
build logically on the existing interest in 'The Creatures House'. Her planning was
developed around the order of play-types she had selected to examine as part of
her participation in the project. This included an early focus on open-ended play,
followed by modelled-play, and concluding with purposefully-framed play. Jeanette invited the five children who had established and maintained the 'The
Creatures House' project to participate in the three consecutively implemented
play-types. These children were Makayla, Lily, Isabelle, Ella and Mason.

4.3 Open-Ended Play

Jeanette gathered the children on the mat whilst they finished their morning tea. They began by discussing creatures. Jeanette asked the children, "Do you think any creatures live in the lake around the island?" One child very excitedly said "a crocodile lives in the lake". Another said "a shark". The children focused on identifying larger creatures rather than smaller bugs and fish in the initial discussion. Jeanette carried on with her questioning and asked, "What will we need to go down to the lake?" The children were well accustomed to going to the lake and knew exactly what was needed. The children put their wet-weather overalls on followed by their gumboots (wellingtons). They also had a red cart with equipment to search for pond life as well as a picnic blanket and other essentials such as a first-aid kit. Once the children were ready to go they made their way down to the lake which was a 10 min walk through a dense 'fairy' garden and farm area. Tracy, one of the researchers attending the walk with Jeanette and the children, described it in her journal as "no ordinary walk. They jumped, ran, skipped, splashed. They took risks—jumping 1 metre off the pier. Their experience was not so much a walk but an adventure".

When they got to the lake Jeanette stopped at a safety sign that said "Danger. Deep Water. Keep Clear". She asked the children what the sign meant. Makayla explained "clear means you can see through it". Meanwhile, Mason thought, "there might be crocodiles". Jeanette agreed with Makayla, saying "clear does mean that, but what this sign also reminds us to do is to be careful and safe around the water." Jeanette and the children moved down to the lake. Jeanette asked the children "How deep is it?" She encouraged the children to walk slowly into the water as it was only a few centimeters deep at the edge. Jeanette provided each child with a magnifying glass and asked them "What can you see?" The children looked around, explored and gradually came together to look more closely together. Jeanette provided the children with some prompts, such as "What can you see when you put your hand in?" "What can you hear?" Lily replied, "I can hear froggies." Jeanette asked "Where might they be?" Lily pointed to a deeper area and they made their way over and looked for a short time. Makayla and Isabelle wanted to go back to the shallow area. The children focused their observations on the plants they could see rather than on finding or identifying any macro-invertebrates as they had discussed at the Centre prior to leaving for their walk. Jeanette observed, "We can see lots of things growing out of the water" and asked, "What might be in the water?" She used further prompts, asking the children "Is there anything swimming in there?" "What's at the bottom?" Ella said "crocodiles" again, but Mason quickly replied, "Crocodiles are not insects" (Fig. 4.3).

As a group the children and Jeanette decided to move to a different location at the lake to try and find more pond life. They were going to walk along the water's edge but Makayla was uncomfortable because she had a hole in her gumboot and her foot was getting wet. Makayla described feeling "scared" several times whilst the children and Jeanette were gathered at the edge of the water. The group then walked a long way around to the pier and the children spent some time leaping

Fig. 4.3 'Crocodiles are not insects' What lives in the lake?

from the pier. The researcher noted in her journal that Makayla's fears seemed to have waned at this point "perhaps because this is what she expects when doing this experience?" Once the children had jumped several times, Jeanette asked "The water is deeper here. Shall we see what we can see in the water?" Makayla, Ella and Jeanette went to the water's edge near the pier to look for more creatures.

Jeanette then gathered the children onto the picnic blanket. She asked them, "Who remembers why we came to the lake? What were we looking for?" Mason replied, "We were looking for bugs". Jeanette then said, "Did we find any?" The children offered a collective "no". So Jeanette asked them another question—"Is there anything else we could use to make it easier to find creatures?"

Makayla: We need to go out into the deep water, but we might sink.
Jeanette: What could we do instead to get to the water?
Lily: We need a boat. Lots of boats.
Ella: Or ride on something.
Mason: I went on holidays once and went on a motorboat.
Jeanette: What could we use to get some of the deep water?
Ella: We could float on the water to collect some? We could use the tree.
Jeanette: These are special trees so we couldn't use those as the birds need them. I think we need to think about how we could get the water for next time so we can see some of the creatures.
Ella: We could get a bowl and scoop it up.
Jeanette: That's a great idea. Ok, we're going to go back now and get warm.

4.4 Modelled-Play

Jeanette's modelled-play experience was implemented 5 days after the open-ended experience. She gathered the children on the mat and asked, "Remember when we went down to the lake to look for creatures?" Was it hard to find them?" Ella

Fig. 4.4 Look and see if you can see anything

replied "They were hiding" and Mason suggested "They were hiding because they were deep in the water". Jeanette had a fish tank on table located near the mat, and so she asked the children "Would it be easier to see the bugs in the tank?" She also pointed out that she had collected some glass jars, nets, trays and magnifying glasses for the children. She said that last time that they had gone to the lake Ella had the idea of collecting some water. Jeanette said "I have got a bucket so we can collect some water as it will be too difficult to take the glass fish tank down to the lake." Jeanette and the children then talked about the different equipment and how it could be used. They discussed how they might use the trays and magnifying glasses to look closely at the bugs. Makayla observed "Some of the bugs might be dead or squished". Jeanette asked her, "How can we tell if they are dead or alive?" Makayla replied, "They won't be moving if they are dead." Jeanette then said that they needed to put on their wet-weather overalls so that they could go to the lake and suggested Makayla wear different gumboots so she didn't get cold feet again.

Like the first time Jeanette and the children made their way down to the lake exploring and discovering along the way. When they got to the edge of the water Jeanette gave each child a net and jar. Together they talked about the holes in the net and how they might catch bugs. Jeanette encouraged the children to go a little deeper into the water and to fill up their jars and then place the water from their jars in the larger bucket. Jeanette observed:

> See how soily this water is? Let's walk really slowly so we can get some cleaner water. We need to go very slowly. When we take it back to the fish tank it needs to be clear so we can see the pond creatures.

Mason and Lily quickly understood what needed to be done. Jeanette encouraged the other children as they were collecting clearer water. Once the children filled the bucket Jeanette asked the children to tip some of the water from the bucket into their trays which were laid out on the picnic blanket. She asked them to "look and see if they could see anything?" They all gathered and began looking at

Fig. 4.5 Ella observing and drawing using the magnifying glass

what they had found in the water (Fig. 4.4). They tended to differentiate between a bug and a plant by whether or not it was moving. When the children found a moving bug they placed it in a jar and watched it swimming and said "let's take the bugs back to the Centre".

4.5 Purposefully-Framed Play

Six days after the modelled-play experience Jeanette gathered the children around a table where she had a fish tank centered in the middle. The fish tank was filled with the water the children had collected from the pond and carried back to the centre in the bucket. Placed around the table was a seat for each child, and at each seat was a tray, a laminated identifying chart, a marker pen, paper and a magnifying glass. Jeanette talked to the children about putting some water from the fish tank into the tray. She initially asked the children "What can you see?" The children were very excited and quickly ascertained that there were many bugs. Jeanette emphasized "We need to remember to 'to look, look and look again'". She asked the children "How is the bug moving?" "How is he swimming?" She said to them "Look at that shape" And asked—"What sort of shape is that?" Ella responded "It's like a spoon because it has a round bit on the bottom." Jeanette prompted Ella and the other children to look at their charts and draw what they saw under the magnifying glass (Fig. 4.5). Ella said "I think it is a beetle". Jeanette then worked with the children attempting to identify the creatures. She continued to use the phrase "look, look and look again" to encourage the children to carefully observe the creatures.

Mason found a snail during his examination of the creatures. Mason and Jeanette then discussed the snail's hard shell. Jeanette also encouraged Mason and

Lily to look at the pond book she had set up on the table to find out more about snails. They found the section on snails and Jeanette asked them "Would you like me to read it to you?"

Jeanette: It says that at first all pond snails look alike as they are very small. When you look carefully you see the shells are very delicate and often like a spiral. Can you see that? Does your snail have a spiral shell?
Mason: No
Jeanette: Have a look, does it have a pointy end?
Lily: It has a sharp bit.
Ella: Look I found a tadpole [consulting the same book as Lily and Mason].
Jeanette: Did you see that in the book that Ella is looking at? There is another pond snail so there are lots of different types of pond snails.

Jeanette continued to engage with the children, exploring other macro-invertebrates and identifying their unique characteristics. To look even more closely at the macro-invertebrates they got the digital microscope and discussed the characteristics of each creature—trying to identify them according to the information in the book. They photographed each creature with the digital microscope and compared them to what they could see in the book.

4.6 Pedagogical Play and Environmental Education

Pedagogical play is understood to encompass a range of play-based activities, including open-ended through to more purposefully-framed play (see Chap. 3). Drawing on a combination of pedagogical play-types forms a more 'intentional' approach to teaching (Epstein 2007) that is arguably of value in environmental education because children require access to content knowledge to engage effectively with environmental concepts, such as biodiversity and animal habitats (Palmer 1993). Jeanette wrote in her journal that she was concerned about the open-ended play experience and the extent to which she would be able to leave this as largely exploratory. Her preference was to engage the children during such play to prompt their thinking and raise possible areas of investigation. She wrote:

My main concern is the open-ended play as I tend to observe children in these experiences and then ask questions to provoke thinking and hopefully extend knowledge and the experience but I know for this research project I must just watch during this session. I do wonder how I will go.

During her interview Jeanette talked about her role in the children's open-ended play in more detail:

I probably talked more than what was recommended for just that play experience because of being at the lake and trying to get them to see. In hindsight I probably would go to another spot next time. I saw frogs there last year so I was hoping to find them again but really I think that there was another place that would have been better to go to that had clearer water to begin with and so perhaps if I had looked at that myself before I went out that is probably one thing I would have done differently. The second time in just providing those materials for them and then taking them out is something that we would do and then the third time pretty much the same thing. Probably what I would do more so is only have two or three children in that final experience when we had five because normally we would only have two or three and then take them.

Whilst Jeanette did talk during the open-ended play, much of her dialogue was in the form of questions which is a feature of the pedagogical approach used in her Centre. Jeanette went on to say that commencing with open-ended play "was a great way to see any prior knowledge the children have". She also said that this is normally where she commences an experience, with open-ended play. She explained:

[Open-ended play] provides an experience where you can see where they are going with it and then being there to see. So sometimes with the experience we may not necessarily do that three times. It may have been brought down to two or sometimes a similar experience we may have taken that moment when they discovered something—found some creatures and then brought them back to class that day, so it would really be how they are going. But asking the questions and trying to get them to respond and look. One of the things when we were on the second walk was what I learned from Tracy [the research assistant]. She was talking about the 'look, look and look again' approach and I have been using that a lot lately because I would say to them 'look and look' but that real 'look, look and look again' to see what they can see I think that is a good one.

Jeanette indicated that the order of play-types she implemented was consistent with her normal approach to supporting children's learning. She suggested that implementing a different order of play-types would have been particularly challenging for her as she used the open-ended as a basis for establishing the children's existing knowledge base and the modelled and purposefully-framed play for building knowledge by connecting more strongly into the biodiversity content she associated with the children's learning. Whilst research suggests intentional teaching is characterised by the use of a range of child-centred to more teacher orientated activities, the exact order or 'blend' of these experiences as a basis for supporting learning tends to remain at the level of teacher discretion (Trawick-Smith 2012, p. 265). In the context of early childhood environmental education, this raises interesting questions because it is not known whether or not children benefit most from the exploratory play first, or would benefit from exposure to the more purposefully-framed play prior to opportunities to engage in the open-ended experiences.

Jeanette's implementation of the play experiences also highlights the need for early childhood environmental education to draw on identified principles for using play-based learning, rather than relying on a conception of play as 'freely chosen' and associated with valuing outdoor activity. Jeanette was asked during her interview what she was hoping the children would learn by the end of the third play-type. Here Jeanette's own disposition towards biophilia was evident, and highlights the important relationship between educator dispositions and the use of play-based learning in early childhood environmental education:

That there are creatures in the water and that they are really tiny and that the creatures they [the children] all thought about were going to be big, they thought that they would see fish, crocodiles and really big things but that there are tiny things that live there and that we need to be careful with them. One died when we tried to get it out last time and since then when we have used it, it is hard to get them out even with a teaspoon and that we do have to be careful with small creatures.

Whilst the open-ended play helped Jeanette ascertain the children's existing knowledge base, it seemed to misconstrue what they believed they would find in the lake (i.e. sharks and crocodiles). It was the later engagement collecting the water and finding the creatures in the water that led the children to an under-standing of the range of creatures that actually lived in the habitat. As such, purposefully-framed play created a context for supporting children's understand-ing of life and therein also supported the development of their own biophilia dispositions alongside Jeanette's. Working towards biophilic dispositions therefore requires using principles of play-based learning that allow biophobic ideas to be explored via the provision of content knowledge with an engaged and supporting educator.

Later in her interview Jeanette was asked why the chart she used with the children during the purposefully-framed play included the heading 'macro-invertebrate' even though she did not explicitly use this term with the children. She went on to say "I didn't use that term but I will do so as we continue on". When asked how far she would go with engaging the content knowledge via the play purposefully-framed play type, she said:

> Probably as a class not so far but individually it will go further depending on the child's interest…I think that we are part of the learning experience so along with the Reggio philosophy we have got the teacher, the environment as well as the experiences, so all of those are important, so if you look at the environment being the teacher, the teacher being the teacher and so on. I think other children they learn so much from each other and you saw that with Lily and Mason and it becomes that scaffold of learning.

An important part of Jeanette's planning in using the three play-types was accessing the outdoors and using the lake as a learning opportunity. During the interview Jeanette was asked to comment on her understanding of the relationship between pedagogical play and the outdoors. This included parental perceptions of the children's experiences on the walks to the lake and activity around the pier:

> **Jeanette:** They are quite mixed. It is interesting we had two of the new parents come yesterday. It was the second time the parents came to the centre and this is week four so they have only been involved in the centre for four weeks and they have been on one other walk and the father is a doctor so I don't know if that has any bearing on it but he was really out of his comfort zone watching Makayla climbing the tree. He stood there and he said how high should we let her climb?
> **Amy:** He said that?
> **Jeanette:** Yes and he said I don't want to step into the boundary of education and I know that you must have a reason for it but he said I find it really uncomfortable and I said to him well safety is a concern and our first concern is always safety and we know that she has climbed a lot of trees before now and that we talk about those safe branches and things like that but having you there in case she falls is a great thing. But it was interesting and on the way down he wanted to lift her down and I said I know but if we can teach her how to get down if we can just guide her, and I said we might need to lift her in parts, but if we can guide her then that is a better thing because then she knows what to do the next time

and he did and at the very end because it was hard to get her leg to the next point he did lift her a bit and that was fine. And he even said afterwards that he found that hard because of that safety thing.

Amy: It was confronting for him?

Jeanette: Yes it was confronting for him and another dad came on a walk and he spent less time involved and it was the first walk that he had been on and he just watched and watched a lot and at the end he said to me "you always said to come on a walk and I thought that we would just walk and I didn't realize how much the children did".

To Jeanette, going on the walk was an important part of the children's learning as it provided access to opportunities for children to be in the environment and to experience nature. Sobel (2008) says that such walks are not 'just a walk':

> If I suggested to my children that we were going on a walk, they complained. However, if I opened with, "Let's go on an adventure," they were much more recruitable. Walks are for adults. You staidly put one foot in front of the other, you chat about boring things with your friends, you wind up at outlooks and say, "Oh what a beautiful view." Snoresville. Adventures mean you don't know what's going to happen when you start out (p. 21).

Jeanette's walks with the children could be understood from an 'adventurous' perspective. However, they also were underpinned by a deliberate consideration of pedagogy because they were not only about the children's outdoor play, but provided a vehicle for engaging Jeanette's own disposition towards biophilia in a way that worked to build the children's knowledge about the environment. Jeanette viewed this as an important aspect of her work:

> Children don't have the same experiences of playing outside as much. Their lives are more controlled I think. We give them those experiences just to play in the environment because they discover as they go. But it is being there to support that discovery so if they find something outside, like we found a dead frog yesterday on the walk, so it is looking at that and talking about that and where could have the frog lived and what might have happened to it.

4.7 Conclusion

The combination of play types implemented by Jeanette was consistent with her normal approach to planning for children's learning. In Jeanette's work there is evidence of the two principles of play-based learning informing her approach to early childhood environmental education. These are:

Principle One:

Valuing different play-types according to their pedagogical potential for engaging with aspects of environmental education

This is seen in Jeanette's use of open-ended exploratory activity to promote initial interest in learning about the macro-invertebrates, and then more adult orientated activity via modelled and purposefully-framed play to further engage children with the content knowledge associated with the initial interest.

Table 4.1 Jeanette's combination of play-types and planned experiences according to the pedagogical value she attributed to the play-types and her associated environmental learning goals

Principle one Pedagogical value Jeanette attributed to play-type	Principle two Combination of play-types Jeanette decided to implement	Planned experience according to play-type	Environmental learning goal associated with play-type
Opportunity to observe children and ask questions to provide insight into existing levels of understanding	Open-ended	Initial discussion: what might we find at the lake? Walk to the lake Search for pond life Concluding discussion: what did we find at the lake?	Exposure to the outdoors, opportunities for risky play and interactions with nature
Raising questions with children that build on existing levels of understanding	Modelled	Initial discussion: remember when we went to the lake? Walk to the lake Collect pond water Examine pond water: what can we see? Take some pond water back to the classroom	Biodiversity: characteristics of macro-invertebrates
Engaging content knowledge to build levels of understanding that contribute to an appreciation of the life needs of other creatures	Purposefully-framed	Fish tank filled with pond water Tray, laminated identification chart, marker pen, paper and magnifying glass Place drops of water into trays Use magnifying glasses: what can you see? Draw what you see Digital microscope and digital photographs Compare digital images to non-fiction text	Biophilic dispositions towards nature

Principle Two:

Creating combinations of play-types that support engagement with different aspects of environmental education

This is evident in Jeanette's combination of open-ended, modelled and purposefully-framed play orientated towards the active building of biophilic dispositions and the acquisition of content knowledge about the macro-invertebrates found living in the lake.

Table 4.1 provides a summary overview of the articulation of the principles to Jeanette's planned play experiences and her environmental learning goals for the children.

References

Cadwell, L. (1997). *Bringing Reggio Emilia home*. New York: Teacher's College Press.

Cornish College (2012a). *Information handbook: Early learning centre*. Retrieved from http://www.cornishcollege.vic.edu.au.

Cornish College (2012b). *Cornish College Educational Philosophy*. Retrieved from http://www.cornishcollege.vic.edu.au/content/educational.

Edwards, C., Gandini, L., & Forman, G. (1998). *The hundred languages of children* (2nd ed.). Greenwich, CT: Ablex.

Epstein, A. S. (2007). *The intentional teacher: Choosing the best strategies for young children's learning*. Washington, D.C.: National Association for the Education of Young Children.

Palmer, J. (1993). Development of concern for the environment and formative experiences of educators. *Journal of Environmental Education, 24*(3), 26–30.

Sobel, D. (2008). *Childhood and nature: Design principles for educators*. Portland, Maine: Stenhouse Publishers.

Trawick-Smith, J. (2012). Teacher-child play interactions to achieve learning outcomes—risks and opportunities. In R. C. Pianta, W. S. Barnett, L. M. Justice, & S. M. Sheridan (Eds.), *Handbook of early childhood education*. USA: Giuldford Publications.

Waller, T., Sandseter, E., Wyver, S., Arlemalm-Hagser, E., & Maynard, T. (2010). The dynamics of early childhood spaces: Opportunities for outdoor play? *European Early Childhood Education Research Journal, 18*(4), 437–445.

Chapter 5
Josh: Small Is Beautiful

Abstract This chapter presents Josh's (an early childhood teacher) and the children's experiences at St Kilda and Balaclava Kindergarten, Melbourne, Australia. The outdoor environment at the kindergarten provided opportunities for children to find and observe living things and explore their habitat. Josh orientated the implementation of his three play-types on an investigation of macro-invertebrate habitats. Josh implemented the play-types in the following order: open-ended play, modelled-play, and then purposefully-framed play. In the open-ended play experience the children explored various habitats with Josh observing the children. During the modelled play Josh modelled finding macro-invertebrates in their habitats, citing their names and characteristics. Josh's purposefully-framed play session began with exploratory learning of the environment, and then matching photographs of the macro-invertebrates with pictures, name, characteristics and habitat. As the children participated in the play-types the level of their biophobic expressions declined and they began to show more biophilic orientated dispositions. Josh was challenged by implementing an open-ended play approach only. He explained he would not normally teach in this manner, instead choosing to follow up the children's emerging interests immediately. However he reflected that he found value in listening to the children's ideas initially without questioning and interacting with the children.

5.1 Josh and the St Kilda and Balaclava Kindergarten

Josh and the children attending his centre are from St Kilda and Balaclava Kindergarten, an inner city childcare centre in Melbourne, Australia. The centre has a strong philosophical approach to teaching environmental education, and has been actively doing so for 10 years. The centre's philosophy states:

> In providing open-ended experiences, children are able to explore, experiment, create, invent and extend upon their innate curiosity using the natural wonders of the world. This helps children in the process of developing knowledge, skills and attitudes necessary to take environmentally responsible action.

The outdoor environment at St Kilda and Balaclava Kindergarten provides opportunities for children to find and observe living things and explore their habitat. Josh orientated the implementation of his three play-types on an investigation of macro-invertebrates habitats for insects, spiders, millipedes and centipedes. The outdoor environment at the centre contained a compost bin, logs, rocks, shrubs and trees, and water tanks. Prior to participating in the project Josh had been focussing on environmental education in his programming with the children for at least 6 months. This included learning about aspects of sustainability, such as recycling, growing food and establishing a composting system.

Josh implemented the play-type combination as open-ended play, followed by modelled-play, and finally purposefully-framed play. Josh began his open-ended play session with five children from the 3-year-old room, including Mitchell, Charlie, Jackson, Netra, Kayne and Anne (Fig. 5.1). One girl, Anne chose not participate in the open-ended play however she participated in the modelled and purposefully-framed play experience.

5.2 Open-Ended Play

Josh asked the children to see what living things they could find in the outdoor area. His goal was to extend and broaden the children's understanding of living things. Throughout the three play sessions he took photographs of the children's findings. The children quickly dispersed throughout the playground as they excitedly suggested where the mini-beasts could be. For example, several children went to the compost bin to see what that could find (Figs. 5.1 and 5.2).

Josh accompanied the children around the yard, and initially the children explored one of the compost bins. One child said he could see spiders, although none were actually evident on this day. This child then asked "Where have all the spiders gone?" demonstrating the child's previous experience with looking in the compost bin. When the compost bin door was first opened another child asked if he could sit on Josh's lap, and Josh reassured the child by saying "it's OK". The first compost bin yielded no living things so the second compost bin was opened. Josh questioned the children about whether there was anything in this bin, while refraining from taking the lead in the exploration. A millipede was found in this bin and Josh asked the children what it was. Netra said "it is a caterpillar", and while the other children knew it was not a caterpillar they did not know its name. At this point Netra began saying in a loud voice "kill it, kill it", while pointing at the millipede. Josh responded by asking Charlie what he thought about killing things. Charlie responded by saying "it is not good". Charlie wanted to touch the

Fig. 5.1 Looking for living things in the outdoor area

Fig. 5.2 Josh and the children looking for living things in the compost bin

millipede and Josh advised just to look at it. Two children picked up leaves and tried to make the millipede move. There was no definitive answer on what it was from the children, and Josh did not tell the children what it was called.

Josh then encouraged the children to look elsewhere prompting "Where else do you think we might find some living things?" Mitchell suggested looking under logs. Stones were turned over first, with children advising their peers to watch their feet. The first stone uncovered some living things, and Netra proceeded to try to kill whatever it was on the ground. Josh advised him to keep his feet back, and a child said it would "snap him". Jackson expressed a desire to hit a snapper bug with a spade, however Josh reminded him of the ethics of caring—"that we look after these bugs". Josh helped the children turn the rocks back over and followed them to the next suggested habitat—some logs under trees. The children assisted each other in turning over the logs, and Josh asked the children if there was anything present. Netra was delighted to find a feather, and asked if he could take it home to which Josh replied "yes". Josh proceeded to accompany the children and helped them to move the logs.

5.3 Modelled-Play

Three weeks after the open-ended play session Josh planned and implemented the modelled-play session. Josh's goals for the session were for the children to think about macro-invertebrates' habitats, their names and characteristics. He also wanted the children to develop a growing appreciation for caring for the mini-beasts by returning them to their habitats after examining them. He led five children—Mitchell, Charlie, Jackson, Kayne and Netra—on a modelled-play experience to achieve these goals. Three of the five children had been involved in the previous open-ended play experience. Mitchell and Netra led the way with enthusiasm to find the living things in the yard. Mitchell declared "Don't kill the bugs, don't kill the bugs". Josh responded to this declaration by gathering the group together and asking Mitchell "how should we treat the bugs?" Mitchell repeated "No killing the bugs, and don't touch them" leading Josh to ask the children "So what could we do? Is it OK if we look at them?" There was general agreement all round regarding this suggestion. Throughout this modelled-play experience Netra, who had expressed interest in killing the bugs in the open-ended play experience, was heard to state on at least six occasions "I'm not killing them".

Josh began this modelled-play session by first going to the compost bins followed by three children holding magnifying glasses. They found a spider's web, no spiders, and some slaters. One child confidently picked up a slater to examine it more closely. There was no response by Josh to this action. With prompting from Josh, the group moved on from the compost bins to look underneath some logs, and whilst making their way to the logs Charlie used his magnifying glass to closely examine the bark of a tree (Fig. 5.3). Upon regrouping the children found a worm and an earwig under the bamboo matting. Netra declared that he wouldn't kill the earwig and Josh thanked him. When the earwig kept crawling into dark areas Josh talked about how it was seeking the dark to be safe. Netra again repeated his declaration that he would not kill it.

Josh's language supported the children's understanding of the habitat associated with the macroinvertebrates. He asked the children questions about the bugs they had found—such as what is its name, characteristics, food and preferred habitat? The children's understanding of the type of habitat the macroinvertebrates preferred was reinforced when the group found plastic under the bamboo matting and Josh asked the children whether the bugs were likely to live in plastic. The children all said "no". Each time Josh uncovered a macroinvertebrate habitat he talked

Fig. 5.3 Charlie using a magnifying glass to examine habitat

about how he was "carefully" returning the cover for the bug, whether it was a log, a rock or the bamboo matting, thus modelling an ethic of care for the living things in the environment, a biophilic disposition.

The discovery of faeces in the yard led the children to speculate that it may have come from a squirrel or even from an elephant. This discussion took a new turn when Netra asked if he could kill the faeces. Josh responded by asking him "is it alive?" Netra said "no" and proceeded to stomp on the faeces. Another sample of faeces was found in the lower branches of a nearby wattle tree, and owing to its size was identified by the children as most likely to be possum faeces. The children agreed that elephant "poo" was probably going to be too large to be found in their yard.

5.4 Purposefully-Framed Play

One week after the modelled-play session Josh began the purposefully-framed experience at the first compost bin with Anne, Netra, Jackson and Jake. Josh began the session by reminding the children of the macroinvertebrates they had found the previous week. He prompted them to remember their search and discoveries. Together Josh and the children examined the layers of rotting matter in the compost bin, found various bugs and discussed the names of the creatures they could see and the food they ate. They found spider's eggs but no spider. Netra asked "where is the spider?" Josh replied that he was unsure. Netra knew that spiders came out of spider eggs, and when he asked if the eggs were edible, Josh replied "we do eat some eggs, but not spider eggs".

The exploration for bugs continued in the yard and each time Josh and the children found a living macroinvertebrate they would discuss its characteristics and try to remember its name. When Anne said she was unable to remember the name of an earwig she said "It was one that Mitchell told us about".

Fig. 5.4 Josh and the children matching found mini-beasts with images placed on a photograph board

In the second compost bin were some tools that did not belong there, including a spade, a ribbon, a piece of plastic and a bucket. The children had established a sense of where things belong in the environment and insisted that these objects be removed. No further bugs were visible so Josh encouraged the children to come and look at photos he had printed of the bugs they had found last week. However the children were still keen to find living things, and proceeded to turn over rocks, and look around the yard. The children were engaged in exploring and discovering more living things in the yard.

Josh began the purposefully-framed play session by stimulating exploratory learning of the environment to find living things. He intended to teach content associated with macro-invertebrates during this session as a follow on from modelled and open-ended play. He planned for the children to match photographs of the macroinvertebrates with pictures (Fig. 5.4), name the macroinvertebrate, and identify its characteristics and habitat. He intended for this knowledge to contribute to the children's appreciation for natural environments, and most importantly to him the development of an ethics of care for living things.

The pictures of spider's eggs that Josh downloaded off the internet were compared with the photograph of spider eggs found in the compost bin the week before during the modelled-play session. The key difference identified by the children was that the spider eggs they had found were "soily" compared to the picture from the internet. However, despite this difference the children did agree that they were the same type of egg. The group agreed to pin the photo of spider eggs next to a photo of two spiders—one male and one female of the brown house spider. The children were interested that the smaller of the spiders was the male, and one child said "girls are small, boys are big". Josh explained how female spiders are bigger than male spiders, and some girls are bigger than boys at kindergarten. Josh then led the children to review what they had seen, and a discussion followed about the differences between millipedes and centipedes, confusion over the names (including calling the centipede a caterpillar) with one child suddenly arriving with a live millipede in his hand. Josh used this opportunity to explain the differences between these macroinvertebrates. At times during this purposefully-framed play experience Anne struggled to find the words to express her understanding, however she was delighted to identify the picture of the worm and she said she knew it was a worm "because it was lumpy". Josh focused the children's attention on the characteristics of each creature—for example the number of legs.

There was some discussion about whether the centipede was dangerous. Netra asserted that all the macroinvertebrates were dangerous, including the centipede, millipede, brown house spider, slug, earwigs, slaters, worms and snail. There was general disagreement about the danger of the macroinvertebrates, although there was agreement that the brown house spider could bite and make one feel unwell.

5.5 Pedagogical Play and Environmental Education

The children responded keenly to investigating what living things were to be found in their yard, and as a result exhibited some knowledge about biodiversity. This included learning the names of the macroinvertebrates, their habitats, and characteristics, and how to care for them in their environment. One week after the purposefully-framed play the children were shown a video of themselves participating in the three play experiences, and asked what they remembered about these experiences. They recalled their knowledge about the body characteristics of spiders, the behaviour of mini-beasts seeking the dark for safety, the names of slaters, spiders, worms, and snappers—however, they still disagreed on the name of the millipede/centipede.

Josh employed several pedagogical strategies during the implementation of the three play-types, including open questioning and speculating during open-ended play, demonstrating in modelled-play, and explaining and engaging in shared thinking and problem solving during purposefully-framed play. Josh's engagement with the children during the implementation of the play-types was a significant support to the children's interactions during each experience. For example, he supported the children to access places in the yard where the macro-invertebrates were to be found, and was available to comfort children when they were uncertain about their findings. This was evident when children expressed fear and some biophobic dispositions towards certain macro-invertebrates. However, given Josh himself held a biophilic disposition he was able to deploy the different play-types to engage the children in rich discussions about the macroinvertebrates. As the children participated in these discussions the level of their biophobic expressions declined and they began to show more biophilic orientated dispositions. This was illustrated by Netra at first wanting to kill and squash the macro-invertebrates and then later being able to say that he would not hurt them.

Josh's teaching style was challenged by implementing an open-ended play approach in one session. He explained that normally he would not teach in this manner, choosing instead to follow up the children's emerging interests within the same day by beginning with open-ended play, then modelled-play and finally purposefully-framed play. This implementation was staggered across a number of weeks, which was not his preferred approach. When Josh reflected upon the open-ended play he said he had found it difficult to refrain from telling the children about the creatures, and as a consequence he realised the value of listening to the children. When asked whether there were any surprises for him by teaching open-ended, modelled and purposefully-framed play Josh replied:

> I was surprised that it was quite hard to stick to them (open-ended questions) because I feel like I use open-ended questioning all the time and a lot of things we do are open-ended but that once I tried to limit what I was saying and the input that you have over the play experience, and I found that really difficult because I always thought I was generally quite quiet and didn't talk a lot but then I thought I do talk all the time when I am working with the children.

Josh realised that his teaching style was interactive and he found it difficult not to influence the children's activity during the open-ended play experience. However each of the three play-types influenced the way he engaged with the children. When using open-ended play Josh restrained himself from giving the children direct content knowledge. As a result, Josh believed he learnt more about the children's thinking and what they were curious about. This included deeper insight into their working knowledge and theories about the macro-invertebrates. During an interview conducted with Josh following the implementation of the three play-types, he was asked whether or not this level of insight was beneficial for supporting the children's learning. He replied:

> Yes trying to not have too much influence over what they were doing… it was an eye opener for me…. it's beneficial for me to recognise that and also it's beneficial just to find out what the children are thinking and what they want to find out and what their knowledge is to stand back a bit.

Josh's insight into the role of open-ended play in learning suggests that such play is a useful starting point for both children and teachers. In Josh's case, the open-ended play was pedagogically valuable when he refrained from too much interaction because it helped him to find out more about what the children were thinking. Trawick-Smith (2012) notes this as a key advantage of the 'trust in play' approach. Interestingly for Josh, this approach was also paired with consideration of the significance of using purposefully-framed play with the children. During his interview Josh was asked to comment on the children's responses to watching the video footage of themselves engaged in the different play-types. Here the children indicated that they preferred the purposefully-framed play when Josh was more fully engaged with them. Josh believed the children preferred this play-type because "they have a real thirst for knowledge, and interest in the subject". Purposefully-framed play enabled him to engage this thirst for knowledge with the children, however, having first implemented open-ended play he was also able to build on what he understood to be the children's current thinking about the macroinvertebrates. In this way interest driven and open-ended play were complemented by Josh's planned experiences and interactions with the intention of building the content knowledge associated with learning about biodiversity.

For Josh, this meant that there was pedagogical value in the combination of the play-types he implemented, commencing with open-ended play, followed by modelled-play and finishing with purposefully-framed play. Josh was aware of raising the children's interest in the topic and hence valued the role of open-ended and modelled-play as a vehicle for establishing this initial interest. Josh felt that if he had begun immediately with purposefully-framed play then only those children with a prior interest in the topic would have continued with the experiences. He believed that beginning with open-ended and modelled-play provided time and opportunity for the children to talk about the concepts and to formulate ideas and questions that were later realised when he explicitly engaged the content knowledge via the purposefully-framed play.

5.6 Chapter Conclusion

Josh believed the combination of all three play-types supported the children's engagement with biodiversity. This was because each play type enabled a range of exploration, reflection and opportunity to participate in discussions that allowed the children to grow their awareness about biodiversity. For Josh, learning and teaching about biodiversity was not so much a matter of choosing to engage with only one type of pedagogical play, rather it meant understanding the benefits of each approach and how they could be combined to realise his goals for the children's learning and their engagement with the natural world. In Josh's work there is evidence of the two principles of play-based learning informing his approach to early childhood environmental education:

Principle One:

Valuing different play-types according to their pedagogical potential for engaging with aspects of environmental education

This is evident in Josh's understanding of the three play-types providing different opportunities for learning that are considered equally valuable. He learned that open-ended play would provide him with insight into the children's thinking, whereas he could use modelled-play to explicitly build children's ethical awareness of how to care for the macro-invertebrates. Purposefully-framed play he viewed as an avenue for addressing the children's 'thirst for knowledge'.

Principle Two:
Creating combinations of play-types that support engagement with different aspects of environmental education

This is seen in Josh's suggestion that modelled and purposefully-framed play extended children's initial open-ended exploratory interests and provided a platform for engaging children in the development of biophilia dispositions.

Table 5.1 Josh's combination of play-types and planned experiences according to the pedagogical value he attributed to the play-types and his associated environmental learning goals

Principle one Pedagogical value Josh attributed to play-type	Principle two Combination of play-types Josh decided to implement	Planned experience according to play-type	Environmental learning goal associated with play-type
Opportunity to observe children and ask questions to provide insight into existing levels of understanding	Open-ended	Explore the outdoor area: where can we find living things? Where else could you look?	Broaden awareness of living things located in the outdoor area Ethics of care for living things
Discussing content to support knowledge building Discussing content to provide a rationale for ethics of care	Modelled	Take the children to the compost bins/logs/matting: what can we see using the magnifying glass? Let's put things back where we find them	Biodiversity content knowledge about macroinvertebrates habitat, names and characteristics Ethics of care for living things
Engaging content knowledge to build levels of understanding that contribute to an appreciation of the life needs of other creatures	Purposefully-framed	Explore the outdoor area: where can find living things	Biodiversity content knowledge about macroinvertebrates' habitat, names and characteristics
		Photograph matching board: what did we find?	Ethics of care for living things Biophilic dispositions towards nature

The use of these principles by Josh in his provision of early childhood environmental education are summarised in Table 5.1.

Reference

Trawick-Smith, J. (2012). Teacher-child play interactions to achieve learning outcomes—risks and opportunities. In R. C. Pianta, W. S. Barnett, L. M. Justice, & S. M. Sheridan (Eds.), *Handbook of early childhood education*. USA: Giuldford Publications.

Chapter 6
Robyn: Worms Underground

Abstract This chapter presents Robyn's (an early childhood teacher) and the childre's experiences. Robyn's kindergarten was located in an outer suburban part of Melbourne, Australia. Robyn focused on worms and making a wormery (a worm farm) owing to the children's interest in worms, and her goals were to provide children with greater understanding of worms and their habitats. Robyn's play order was purposefully framed play, modelled play and open-ended play. Robyn brought in a large clod of soil rich in worms for the children to explore. She initiated purposefully framed play asking children in-depth questions about worms, used correct terminology and built a worm farm with the children. She used nonfiction books and scientific tools to enhance the children's learning. During the modelled play children made their own worm farms under guidance from Robyn, and in the open-ended play they were given free rein to make their worm farm. This led to some of worms being drowned by the children using too much water, leading Robyn to step in to save the worms. Robyn identified that normally she would use all three strategies in the same play session rather than in isolation or separately. This articulation of the combined play types to environmental education is important for early childhood education as educators move in and out of teaching strategies depending on the children's cues, their interests and the intent of teaching.

6.1 Introducing Robyn and Hallam Kindergarten

Robyn's kindergarten was located in an outer suburban part of Victoria, Australia. There were many children from diverse cultural backgrounds and from lower to mid socioeconomic circumstances in Robyn's kindergarten. Robyn valued outdoor play for young children because of the opportunities she perceived it provided for children's learning about the environment. She was particularly interested in supporting children to be respectful of other living creatures and viewed outdoor play as an avenue for achieving this goal. Robyn believed in the importance of

A. Cutter-Mackenzie et al., *Young Children's Play and Environmental Education*
in Early Childhood Education, SpringerBriefs in Education,
DOI: 10.1007/978-3-319-03740-0_6, © The Author(s) 2014

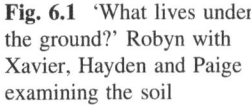
Fig. 6.1 'What lives under the ground?' Robyn with Xavier, Hayden and Paige examining the soil

what she called 'direct' teaching as well as open-ended play because she felt the cultural experiences of the children she worked with warranted strong interactions and engagements to foster learning.

During her participation in the project Robyn decided to focus on worms and making a wormery (a worm farm). Robyn mentioned that the children were already interested in worms, and said that "every time they found one they were extremely excited and then they just waned off". Taking this reflection into consideration, Robyn felt that the children's interest needed to be extended to provide "greater understanding about worms" not just encouraging basic "observation skills". Consequently, Robyn decided to commence with purposefully-framed play, then implemented a modelled-play experience and ended with open-ended play. Robyn intended for the children to learn how the worms moved, what they needed to survive and the nature of their habitat.

Three boys (Hayden, Xavier and Sam) and two girls (Paige and Tahima) participated in the three play experiences. While many children had parental permission to participate (and also provided self-consent), these five children were chosen by Robyn either because they were particularly interested in worms, or because they were "always playing in the soil". Robyn's approach to making a wormery was resourceful and constructive. Robyn particularly wanted the children to appreciate that worms live in the natural world "under the ground" not just in an artificial, plastic worm farm. This meant the basis of the physical setup for her experiences was a large grassy topped clump of soil freshly dug up from her paddock at home because she considered the soil at the centre to be "not good for digging" and not "bug rich" (Fig. 6.1).

6.2 Purposefully-Framed Play

While the rest of the children in the centre continued to play inside with other educators, Robyn gathered the five children to work outside on "something special", walking past the native grasses to sit at a small table with bench seats under

Fig. 6.2 Xavier showing
Robyn a worm in the book
similar to those in the
wormery

the eucalyptus trees. A large plod of grass covered soil nearly covered the whole
table top and the children were invited to sit and discuss "what lives under the
ground?" The first question the children wanted to know was "where did you get
this ground from?" showing an immediate awareness that this piece of earth was
different to the soil at their centre. Robyn explained that the soil came from her
paddock at home and the children helped to lift the heavy piece of rich earth upside
down, so the soil was exposed under the grass to aid their investigation. The
children were instantly engaged, excited and interested in finding "what was under
the ground", whilst Robyn asked several questions—"What do you think might
live under the ground?" and "How will you know if it's a worm you've found
under the ground?" Correct scientific terminology was used by Robyn and the
concept of habitats was explained. There was a focus on asking biophilia orientated questions such as, "Are worms safe for us to touch?" and "What do you
think worms do under the ground?" Robyn's questions punctuated the children's
activity, with the children responding, and saying "I found a big giant worm" and
"I found a baby worm because he is little", as they carefully sorted through the
soil, finding and picking up worms.

The children were then invited to build a worm house together. Robyn asked the
children what they would need to make the worm house similar to the soil they had
just examined. Next to the table where they were playing with the soil Robyn had
placed a number of containers with sand, additional soil, compost/food and newspaper, along with small shovels for the children to collect materials to make the
wormery. The children were again invited to collect small amounts of these materials
to pour into the large glass fronted wormery in layers, before covering it in black
paper to simulate the darkness of under the ground (see Fig. 6.2). This session was
45 min long, requiring sustained periods of concentration on the children's part.
Perhaps because of this, Tahima started to wander away to the swings and tended to
come back and forth, while Xavier was more involved in the 'Earth Worm' picture
story books nearby than the actual worms in the soil, exclaiming to Robyn, "Look, a
worm" pointing to the pictures.

For the other children water was added to the mix of materials to make the
habitat for the worms more life-like. Robyn continued to pose questions, such as,
"Do worms like it wet or dry?" Hayden speculated, "Wet... so worms can live

underground at the beach too!" Sam suggested the wormery should stay outside "because it's much cooler outside, and worms like it cold!!" These comments seemed to demonstrate the children's developing understanding about the nature of the habitat needed by the worms.

6.3 Modelled-Play

The following week, the same five children—Hayden, Paige, Sam, Xavier and Tahima -were invited to look at worms again, but this time the session was held inside. At the beginning of this experience, the children gathered around Robyn for a discussion about "what do we need to do to make this wormery like under the ground for the worms?" This time the children were invited to make their own individual wormeries using a plastic bottle inside another half bottle, with soil and other materials inserted in the space between, then eventually covered in black paper to make it dark "like under the ground". They talked about what the worms needed to live happily, and how the wormery would need drainage holes. The children appeared engaged and interested in the experience and were eager to start.

Back at the table was another huge plod of grassy earth waiting for the children to discover "what lay under the ground". On the floor, surrounding the table were the same materials the children had used to make the large wormery during their purposefully-planned experience including, containers of sand, soil, newspaper with spoons and small cups. Additional resources including non-fiction books about worms and magnifying glasses were also provided. This time, the children knew immediately what to do, and collectively started to lift the heavy patch upside down with exclamations of "I found a wriggling worm" and shortly afterwards, "Look everyone, I found a worm… He's with his family." This statement led Robyn to state that "He belongs under the ground!" which triggered the children to start making their own individual wormeries to remedy the situation of homeless worms.

This session lasted for 65 min and was characterised by the children's sense of innovation as they made their own wormeries. The children readily drew on the resources available, selecting what they believed was necessary to create their worm farms. All of the children took care and time to compact soil, sand and compost (for food) into their bottles, with added "decorations" on top to make their wormeries look like the "grassy patch". The children had understood the concept that worms live "under the ground", and that they needed to replicate that habitat. Occasionally there were issues of worm ownership, with Xavier saying, "That's my worm" while Hayden asked, "Please can I have one of your worms?" Overall, the worms were carefully collected, examined, shared and located in each child's own wormery (See Fig. 6.3). During this experience some children chose to use the spoons provided for their investigation and collection of worms while others used only their bare, muddy hands.

During this experience, Robyn sat at the table with the children asking questions such as, "Are you trying to find out what the worms like in their house?"

Fig. 6.3 Robyn modelling
the making of a womery with
Hayden and Xavier

The children continued to experiment and investigate, answering Robyn's questions whilst they engaged in the experience. While Robyn was clearly involved and interested in their activity, she was not directing the children, instead adding comments and questions that deepened the children's inquiry into worm life. For example, at one stage, Robyn asked, "Do worms have babies?" to which Sam answered quickly, "Yes, because they have families!" Another question posed by Robyn about how worms eat, was answered by Hayden who said, "Worms are small, so they eat small stuff like compost." Robyn's questions and comments appeared to provoke the children's scientific thinking. This notion was especially evident when Robyn commented about one worm's travels, saying "Look, he went all the way down to the bottom. I wonder how he did that. How does a worm get all the way through the soil?" The children's answers show the collective peer thinking and learning occurring throughout this play with Hayden answering:

> A worm is very small, and doesn't have any hands, so he pushes his way through with his
> head.

Sam listening close by added, "He has very strong brains." It was at this stage that Robyn mentioned to the children, "If the worms are going to live in there, they'll need it to be a little bit wet!" This comment had unforeseen ramifications for what was to occur during the children's participation in the open-ended play experience a week later.

6.4   Open-Ended Play

Another week passed before the open-ended play type session was planned for the children. Once again, the play was set up inside the preschool, with Robyn inviting the children to make their own wormeries centred on the tray of soil, materials and

Fig. 6.4 Hayden and Paige
making their worm farms

resources they had used before, and then she stepped away from the group. This
time, spray bottles of water, plus torches were added to the experience in an
attempt to provide an additional mode of investigation of worms in the earth
(Fig. 6.4) Another addition was a raised, clear tray of soil for children to be able to
see worm action from underneath.

Raised children's voices were heard back and forth across the table with
comments ranging from "There's no worms here!" to "Found one" and then,
"Wow, look at this worm!" The children flicked the torches on and off and
experimented with the magnifying glasses. The spray bottles of water were con-
stantly in use, but it was not the occasional squirt of water to dampen the soil that
was happening—instead large muddy puddles of water were appearing over the
table top and on the floor. At one stage, worms were being flicked into children's
faces. Some of the children tried to rectify the issues saying to each other, "Don't
take another worm!" Eventually it was the appearance of the drowning worms that
drove Robyn to return to the table, where she quickly reverted to commenting and
questioning the children in her effort to save the worms. Robyn asked the children
with a horrified look on her face, "Do they look like they are moving in there? Can
worms swim?" and then she said, "Do you think the worms would like the water
or the soil to live in?" In Robyn's presence the investigation once again became
richer, with children commenting "I found something too small to be a worm, and
too small to be a caterpillar." Their behaviour also became more respectful
towards the worms who were less flooded and more gently treated. After an
extended period of sorting and collecting worms, the children started to make
wormeries again as suggested at the beginning of the session. Robyn reminded the
children that "he's got soft skin, if you touch him too hard he'll be hurt."
Conversations about other minibeasts added another dimension, with Tahima
suggesting that she had "just touched a little green worm, but I dropped it and it
runned away." Robyn enquired, "Oh, could you see its legs?" which in turn
developed into further discussion about the attributes of worms.

Eventually, Paige, Tahima and Xavier decided they had had enough play with the worms, and wandered away from the table. The final two boys, Hayden and Sam, who were left at the table, appeared to be totally immersed in the sensory elements of the wormery-making play, including making up musical rhymes in the mud. Robyn wondered aloud after an hour, "How will you know when it's finished?" which was answered by one of the boys saying, "I'll tell you, because I love building it." Robyn replied, "I'm sure the worms appreciate you being gentle!" Finally Robyn suggested to Hayden and Sam they had been working there for such a long time, that perhaps they could come back later and check how the worms were in their wormeries? The boys thought this was a great idea, and decided, "We'll come back later!"

6.5 Pedagogical Play and Environmental Education

Robyn reflected on the relationship between the three play-types, her planning for the children's learning, and the learning she believed had occurred during the children's experiences. Although Robyn initially planned to have all the play experiences outside, she felt it was difficult to "contain the children" and that being outside was more of a distraction for the children who at times wandered away from the purposefully-framed play experience. For the subsequent play types, Robyn arranged these to be set up inside while the other children were playing outside:

> I thought that being outside would make it more natural, more getting into the experience more readily, but to me it was more of a distraction than an enhancement.

Robyn was asked if she thought her teaching style influenced the way she interacted with the children differently in each play type. She commented that as a teacher she considers herself to be "quite directive", she was surprised how difficult it was "to separate the mix of all three different play-types", and to manipulate the teaching in the way she had intended:

> You can see I was trying to pull it back to a more purposefully-framed experience. And it was interesting because I find that quite often as a teacher I am quite directive and frame some of the experiences quite firmly, and yet when I tried to make it definitely a purposefully framed thing and when it went into a slightly different direction I found it really hard to bring it back to that.

On reflection, Robyn concluded that as a teacher "you just automatically jump in and out of the different play-types based on what you need." This was supported in an entry recorded in her journal where she wrote "teachers blend teaching strategies to help children work constructively towards goals." Robyn's claim of "blending" play-types acknowledges existing ideas regarding intentional teaching in early childhood education, including the argument that children benefit in terms of knowledge construction from experiences that contain a balance of adult-initiated and child-centred activity (Epstein 2007; Wood 2013). The articulation of the combined play-types to environmental education is important for early childhood education as it suggests using a range of pedagogical play for supporting children's engagement with environmental knowledge, rather than relying solely on one play-type to enable learning.

One of the most significant concerns to emerge during Robyn's implementation of the three play-types was that environmental education for young children needs engaged educator support and not just self-discovery through open-ended play experiences. It seemed from Robyn's reflections and journal writing that this engagement included using a combination of the play-types to support children's learning. For example, it was during the implementation of the open-ended play where the children began to over-water the worms that Robyn said, "I couldn't help myself" and felt she had to intervene during the play when "creatures were being hurt". It was clear in this situation that open-ended play suggested opportunities for the children to mistreat the worms in a way not consistent with the messages of nurturing and caring for living things. Robyn was asked if she thought the children's learning about the worms would have been as engaged if only open-ended play had been used:

> No…when we had open-ended play here I was trying to stay out of it, but when the worms were getting stretched and hammered and drowned, I just couldn't do that.

Pearson and Degotardi (2009) suggest that children's learning about sustainability in early childhood is important because research shows that biophilic attitudes towards the environment are formed during the early years. These attitudes may not develop in a meaningful way in situations where open-ended play lessens the access children have to adults for supporting their interactions with other living things. For example, the children in this study tended to default to harmful behaviour towards the worms when there was less adult support available to them during the open-ended play experience than they had exhibited during the purposefully-framed and modelled play activities. This finding is interesting because Robyn felt she and the children had already done a "fair bit of work" about respecting all living things prior to their engagement in the final open-ended experience. However, Robyn concluded that a biophilic attitude takes time to establish:

> Just that the bugs we don't have to harm them that they are living things and they breed like us. 'They don't like to be hurt, do you like to be hurt?' 'They've got soft bodies, do you like your body to be hurt?' 'They don't have big bones like we do'. I just try to create an understanding and empathy I suppose about creatures, and I guess that comes from my belief that we are all entitled to live. Even with spiders we will put them in a container so we can watch them. We have a lot of Red Backs [small highly poisonous spiders] around here so even when we remove them from the environment, we do it so the children don't see how they are removed, so we often catch them in a container and then they are removed later. But other spiders we will put out into the garden because they've got a job to do. So I think all of those attitudes build up over time it's not just something that you teach in one hit.

Robyn also suggested that it can be difficult to support biophilia when there is minimal support for this type of environmental learning from the children's home, and said, "I don't think they spend a lot of time outside in the outdoor environment as a family." Robyn was particularly clear about her attitude and believed that "a bit of soil doesn't hurt us" in relation to encouraging the children to experience the reality of worms and the soil without the use of "latex gloves". At times throughout the worm sessions, there were comments from some of the children demonstrating their anxiety about "the mess" being created in the play, with one child warning others "don't touch, it's dirty". Such comments are somewhat typical signs of the formation of early biophobic attitudes towards nature.

The conversation about the range of play-behaviours exhibited by the children during their participation in the the open-ended and purposefully-framed play triggered some interesting comments from Robyn regarding the rhetoric associated with open-ended play and its role in young children's learning. Robyn was asked if she thought it was still the dominant pedagogy used in early childhood settings:

> I think there are some places where perhaps it is, but I am not fully sure that everyone knows what open-ended means. It is not my predominant thing. I would probably say I use more supported and modelled and guided play and I probably use guided more than purposeful…
>
> I think that if you don't have a purpose for what you put in the room, my belief is that everything you model and say and do from the minute you walk in the door sets the scene and influences what happens in the day.

The role of the adult in children's learning is a significant aspect of contemporary perspectives on intentional teaching (Fleer 2010; I. Siraj-Blatchford 2009). This appeared to be the case for Robyn, whose implementation of the purposefully-framed and modelled-play, included using in-depth questioning, the use of correct terminology and extensive visual cues such as the clump of earth, non-fiction books and scientific tools. These pedagogical strategies supported the children's engaged and extended play with complex biodiversity concepts, such that children participated in the experiences for up to sixty minutes. For example, during the purposefully-framed play experience the children talked about what the worms looked like, what they ate and what constituted a worm family. In the modelled-play experience, they discussed what worms need for their habitat and how such a habitat could be re-created in a wormery. The combination of these strategies possibly contributed to the type of "deep learning" Littledyke and McCrea (2009) claim is necessary for young children to engage with scientific content (p.43). Robyn reflected on the nature of this "deep learning", indicating that she "would never have anticipated the length of engaged time that evolved" during the purposefully-framed play, and that she was surprised that even more information emerged during the modelled-play type in the subsequent session. It was noted that Robyn's provision of multiple "visual cues" together with asking in-depth questions were highly successful pedagogical aids that assisted the children's extensive engagement in the experiences:

> That's probably my style that I try to include some of the information into the question, provocations for them but try to only give it in small amounts until it seems that they are ready for the next bit, so I think with the saddle someone noticed that he had a 'band-aid' [sticking plaster] so it was giving him the actual name [by mentioning the saddle].

Table 6.1 Robyn's combination of play-types and planned experiences according to the pedagogical value she attributed to the play-types and her associated environmental learning goals

Principle one Pedagogical value Robyn attributed to play-type	Principle two Combination of play-types Robyn decided to implement	Planned experience according to play-type	Environmental learning goal associated with play-type
Supports strong interactions and engagements for fostering learning	Purposefully-framed	Large clump of soil: what lives under the ground? How do you know? Why do you think worms live under ground? Are they safe for us to touch? Use correct terminology: worm, soil, saddle, castings Build a large worm together Cross reference worm farm with non-fiction books	Move children beyond 'just looking' at worms towards acquiring content knowledge about worms: habitat, characteristics
Allows teacher to 'set the scene' influencing the type of learning that occurs	Modelled	Materials for making individual worm farms How can we make these farms look like under the ground? Look at the clump of soil Show how to use magnifying glass to examine worms Provide non-fiction books about worms	Content knowledge about habitat: moisture, drainage Respect for living things
Exploration and access to nature (opportunity to get dirty)	Open-ended	Provide materials for making individual worm farms Spray bottles of water for moistening soil	Biophilic dispositions

6.6 Conclusion

Robyn's implementation of the three play-types illustrated the importance of adult interactions during play to support the children's engagement with the biodiversity content forming the basis of the experiences. In Robyn's situation adult interaction involved questioning, providing information and prompting the children's understanding about the worms and their habitats. During the open-ended play experience, the absence of these interactions meant that the children began to engage in what was undoubtedly satisfying exploratory play for them—but considerably less so for the worms. This led Robyn to reflect on the extent to which open-ended play alone would be a satisfactory pedagogical approach in relation to early childhood environmental education. This reflection is consistent with contemporary research regarding the role of open-ended play in early childhood education in relation to children's engagement with content knowledge—particularly where this has shown that children are more likely to build conceptual knowledge via interactions with adults. Importantly for environmental education, such child–adult interactions seem to be important for addressing early signs of biophobia amongst children and supporting the development of biophilic dispositions.

Principle One:

Valuing different play-types according to their pedagogical potential for engaging with aspects of environmental education.

This is evidenced in Robyn's understanding that whilst open-ended play supports children to become comfortable with aspects of nature (such as playing with soil), that modelled and purposefully-framed play provides a basis for learning why it is important to be respectful towards other living creatures.

Principle Two:

Creating combinations of play-types that support engagement with different aspects of environmental education.

This is reflected in Robyn's ideas about "just jumping in and out of the different play-types based on what you need" to support learning. Sometimes Robyn blended play-types within one type (for example stepping in with modelled play when she saw the worms being harmed by the children during the open-ended experience).

The use of these principles by Robyn in her provision of early childhood environmental education are summarised in Table 6.1.

References

Epstein, A. S. (2007). *The intentional teacher: Choosing the best strategies for young children's learning*. Washington, D.C.: National Association for the Education of Young Children.

Fleer, M. (2010). *Early learning and development: Cultural-historical concepts in play*. Melbourne: Cambridge University Press.

Littledyke, M., & McCrea, N. (2009). Starting sustainability early: Young children exploring people and places. In N. Taylor & C. Eames (Eds.), *Education for sustainability in the primary curriculum: A guide for teachers* (pp. 39–57). South Yarra, Vic.: Palgrave Macmillan.

Pearson, E., & Degotardi, S. (2009). Education for sustainable development in early childhood education: A global solution to local concerns? *International Journal of Early Childhood, 41*(2), 97–111.

Siraj-Blatchford, I. (2009). Conceptualising progression in the pedagogy of play and sustained shared thinking in early childhood education: A vygotskian perspective. *Educational and Child Psychology, 26*(2), 77–89.

Wood, E. (2013). *Play, learning and the early childhood curriculum*, (3rd ed.). London: Sage Publications.

Chapter 7
A Challenge Reconsidered: Play-Based Learning in Early Childhood Environmental Education

Abstract In this chapter, the authors discuss the two principles that emerged from this research project, and that can be applied for play-based learning in early childhood environmental education. These principles are (1) Valuing different play-types according to their pedagogical potential for engaging with aspects of environmental education; and (2) Creating combinations of play-types that support engagement with different aspects of environmental education. These two principles go beyond the traditional thinking of learning 'naturally' through play. This is because the principles allow educators to identify pedagogical value associated with a play type and to combine this with other play types to achieve environmental learning goals with children. Simply providing children with access to open-ended play in an outdoor setting is insufficient to support environmental learning. Environmental learning in the early years needs to provide children with opportunities for acquiring content knowledge that allow them to build understandings about their world and develop biophilic dispositions toward nature. This is a necessary basis for engaging children in discussion about the need for sustainability and sustainable actions in their own lives and communities.

7.1 Introduction

The opening vignette for this book described four children happily playing with some materials placed in a wading pool at their kindergarten. It was noted that the materials had been selected to help the children learn about the biodiversity associated with their local beach. Whilst the children initiated a play-script drawing on their knowledge of SpongeBob Squarepants, their teacher (Seth) approvingly noted the positive nature of their social interactions. Such play-experiences are not uncommon in early childhood education, and are believed to help children learn to respect other living things and to develop an ecocentric appreciation for the environment. Our argument, in the context of emerging research into the use of pedagogical play in early childhood (see Chaps. 2 and 3),

A. Cutter-Mackenzie et al., *Young Children's Play and Environmental Education in Early Childhood Education*, SpringerBriefs in Education, DOI: 10.1007/978-3-319-03740-0_7, © The Author(s) 2014

and the debates associated with environmental education (see Chap. 3), is that such activity is not enough to support children's environmental learning.

In this book we have considered the history of play-based learning and seen that critiques of the historically valued use of open-ended play have resulted in new conceptions of pedagogical play that value the role of adults in children's activity. Variously conceptualised as intentional teaching (Epstein 2007), sustained shared thinking (Siraj-Blatchford 2009), integrated pedagogies (Wood 2013), pedagogical activity (Dockett 2011) and inter-contextuality (Fleer 2011), these positions suggest that adult-child interactions during play support conceptual learning. In this book we have drawn on three different play-types to show how such interactions can be realised by teachers engaging children in environmental education. Broadly, the play-types used in this book mirror those established by Trawick-Smith (2012) as the 'trust in play', 'facilitate play' and 'learn and teach play' approaches, and include open-ended play, modelled play and purposefully-framed play.

In Chap. 4 we saw how Jeanette used open-ended, modelled and purposefully-framed play to foster children's learning about pond life. Jeanette suggested that the combination of play-types she used was a significant aspect of how she understood the relationship between play-based learning and environmental education. This was because Jeanette believed that it was important to provide children with opportunities to explore and examine the environment via open-ended play prior to participating in more adult-initiated experiences such as modelled and purposefully-framed play. However, Jeanette, like Robyn, noted that open-ended play alone was insufficient for dispelling misconceptions the children held about the likely creatures to live in the lake and the promotion of biophilic dispositions towards nature. Here, both Jeanette and Robyn valued purposefully-framed and modelled play as play-types that enabled them to create direct relationships between the children's open-ended experiences and the range of content knowledge that both dispelled misconceptions and helped to reduce biophobic dispositions. This was clearly the case when Robyn intervened in the children's open-ended play to protect the worms from being "overwatered, stretched and hammered" by the children.

Chapter 5 highlighted Josh's understanding of the play-types as providing the children with different but equally valuable opportunities for engaging and supporting learning. Josh suggested that open-ended play was useful because it allowed him to observe and listen to the children in ways that alerted him to what they already knew and understood about macroinvertebrates. Modelled-play was perceived as providing a prime opportunity for demonstrating ethical ways of engaging with macroinvertebrates and for challenging biophobic tendencies. Meanwhile, purposefully-framed play was viewed as significant because it allowed Josh to support the children's interests and gave them access to information that further supported their developing ethical perspectives. For example, learning that insects prefer dark spaces helped the children decide to return them to their habitat after they had finished looking at them. This stance on purposefully-framed play was echoed by Jeanette and Robyn, each of whom reflected on the extent to which actively engaging the children in content knowledge about biodiversity later prompted more ethical and respectful interactions with the environment on the children's behalf.

Chapter 6 focussed on Robyn's implementation of the three play-types commencing with purposefully-framed play, then modelled play and ending with open-ended play. Here, Robyn argued that a fundamental environmental concern of hers was that the children would learn to respect the life needs and rights of all creatures—including worms. Robyn was also keen for the children to understand worm habitats, and created an ingenious series of experiences where children were able to see, touch and explore a large clump of soil from above, the side and from underneath prior to constructing their own worm farms. Purposefully-framed play mattered because being able to talk to the children about worm habitats provided a basis for engaging in conversation about the 'rights' of the worms during open-ended play. Thus, like Josh, ethical and biophilic dispositions towards nature were enabled by Robyn when she used content knowledge as a basis for talking with the children about the characteristics of the worms and their preferred habitats.

7.2   Two Principles for Using Play-Based Learning in Early Childhood Environmental Education

The perspectives held by Jeanette, Josh and Robyn suggest two principles for using play-based learning in early childhood environmental education. As noted in Chap. 1 the first principle is concerned with the pedagogical *value* associated with each play type, whilst the second principle is concerned with the *combination* of play-types educators use to achieve different environmental learning goals. These principles were represented in Tables 4.1, 5.1 and 6.1 at the end of Jeanette, Josh and Robyn's chapters in relation to the experiences they planned and implemented for the children according to the goals they held for the children's environmental learning outcomes. In summary, the principles may be understood as:

Principle One

Valuing different play-types according to their pedagogical potential for engaging with aspects of environmental education.

This principle is based on the idea that no one play type is more valuable than another. Each play-type offers particular experiences and opportunities that help teachers to think about children's learning, and therefore their approach to teaching environmental education. Open-ended play can support the exploration that exposes children to nature and biodiversity. As Jeanette and Josh noted, open-ended play can also provide teachers insight into children's current modes of thinking. Modelled-play can promote opportunities for teachers to illustrate respectful relationships with the environment. Purposefully-framed play can enable access to content knowledge that further promotes biophilic dispositions. Valuing play-types for the different pedagogical potential they provide teachers means open-ended play does not have to be promoted over and above the other play types. This is consistent with contemporary research regarding the use of 'integrated' play-based pedagogies (Wood 2013) because it highlights the unique pedagogical value of each play-type.

Principle Two

Creating combinations of play-types that support engagement with different aspects of environmental education.

The play-types do not need to be fixed in a given sequence to usefully support teachers and children engaged in environmental education. Instead, the play-types can be used in flexible combinations to meet teacher goals for children's learning according to the value they attribute to a play-type. For example, Josh valued open-ended play because it provided him with insight into the children's current levels of thinking and understanding. In turn he appreciated purposefully-framed play because it helped him to support children's ethical encounters with the macroinvertebrates. Modelled play was valued because it allowed Josh to share his own biophilic dispositions. If Josh's goal was to promote biophilic dispositions with the children he might deliberately begin with modelled play, prior to moving to purposefully-framed play and then open-ended play. Conversely, if Josh was more focused on engaging the children in learning about the habitats of the macroinvertebrates he might commence with open-ended play so that he could gain insight into their current thinking and then use this as a basis for planning some purposefully-framed activity. Here there is no need to use all three play-types in a particular order. Rather iterations of play-types can be selected and sequenced by teachers to achieve their environmental learning goals for children, whether these are orientated towards supporting biophilic dispositions, the acquisition of content knowledge associated with biodiversity, developing eco-centric perspectives on the environment or engaging with sustainability.

7.3 Re-considering Play-Based Learning in Early Childhood Environmental Education?

In the introduction to the book we suggested the two principles for using play-based learning in early childhood environmental education could be used to re-consider the provision of Seth's initial experience for the children gathered around the wading pool. The activity as first described could be taken as an open-ended play experience. If like Josh, Seth valued open-ended play for the potential insight it was likely to give him into the children's thinking he may have noticed that the children related the sea creatures to SpongeBob Squarepants and his sea star sidekick Patrick. This could have alerted Seth to the fact that the children had some pre-existing knowledge of sponges and sea stars. Seth's next step could be to consider what he valued as the pedagogical potential for both modelled and purposefully-framed

Table 7.1 Seth's re-considered approach to early childhood environmental education using the two principles of play based learning

Principle one	Principle two		
Pedagogical value Seth attributes to play-type	Combination of play-types Seth decides to implement	Planned experience according to play-type	Environmental learning goal associated with play-type
Provides insight into children's existing knowledge base as informed by their media viewing	Open-ended	Place seaweed, sea stars and plastic sea animals in wading pool	Identifying sea creatures
Engaging children in content knowledge to expand existing media-informed knowledge	Purposefully-framed	Place seaweed, sea stars and plastic sea animals in wading pool Locate books and/or iPad with information/ video about characteristics of seaweed and sea stars near wading pool/ Watch Sponge Bob Squarepants Read, watch, share and discuss video and information	Content knowledge about biodiversity: characteristics of seaweed, sea stars, octopus, fish
Opportunity for modelling biophilic dispositions	Modelled	Visit the beach Search for sea creatures Return creature to habitats Photograph habitats Create photographic habitats for plastic creatures and/or characters from Sponge Bob Squarepants	Respecting living things Content knowledge about the relationship between creatures and habitat

play. If purposefully-framed play was perceived by Seth as providing opportunities for engaging children in content knowledge about sea creatures he may plan to build on what they already recognised as the difference between the sponge and the sea star. Several possibilities are evident here—perhaps Seth would choose a website or some video footage to show the children using an iPad; perhaps he would source some books and engage the children in discussion about the different features on each creature. Maybe they would watch an episode of Sponge Bob Squarepants and discuss the various creatures represented as characters in the program. Having done this, Seth might be interested in expanding the children's awareness of the habitats associated with each creature. Using modelled play he may plan to re-visit the beach showing the children how to search for the creatures and being careful to return what he found to the correct habitat. Photographs could be taken of the different habitats and on return to the centre the children could use these images to create 'homes' for the plastic sea animals used in the initial play-experience. Alternatively, the children may be invited to create or paint appropriate habitats for the different characters from the Sponge Bob Squarepants program.

Re-thinking Seth's provision of early childhood environmental education using the two principles of play-based learning highlights how considering the *value* (Principle One) associated with a play-type informs why a teacher might decide to use that type, and consequently the *combination* (Principle Two) of types a teacher might decide to implement to achieve particular environmental learning goals. This is because we can see that Seth's decision to use modelled play and then purposefully framed after the open-ended play is based on providing the children with access to the some content information (modelled) and the building on this information to extend their understandings of habitat (purposefully-framed). The two principles of play-based learning articulate with each other to provide Seth with a framework for approaching early childhood environmental education that allows him to move beyond simply providing experiences to engaging in learning about biodiversity with the children.

Like the examples provided for Jeanette, Josh and Robyn (Tables 4.1, 5.1 and 6.1) it is possible to illustrate Seth's re-considered approach to early childhood environmental education using the two principles of play based learning (Table 7.1).

7.4 Conclusion

In the opening chapter of this book we suggested that traditional play-based practices posed a challenge for early childhood environmental education. This is because simply providing children with access to open-ended play, the outdoors and nature is not enough to support environmental learning. Environmental learning in the early years needs to provide children with opportunities for acquiring content knowledge that allow them to build understandings about their world and develop biophilic dispositions toward nature. This is a necessary basis for engaging children in discussion about the need for sustainability and sustainable actions in their own lives and communities. Furthermore, research in play-based learning over the last decade has suggested multiple ways of thinking about how adults can most effectively engage with children during play-based activities to promote learning. Our work with educators and children suggests that the two principles of play-based learning we have identified in this book can be readily articulated to the provision of early childhood environmental education. This is because the principles allow educators to identify the pedagogical *value* they associate with a play-type and to *combine* this with other play-types in order to achieve environmental learning goals with children. Whilst there will always be challenges associated with how best to engage young children in environmental education, these two principles go some way to providing educators such as Seth with a starting point for more readily integrating such education into the early years. In this way, children are able to transcend traditional notions of environmental learning (such as swirling seaweed or stretching worms) to participating instead in deeply rich play-based early childhood environmental education.

References

Dockett, S. (2011). The challenge of play for early childhood education. In S. Rogers (Ed.), *Rethinking play and pedagogy in early childhood education: Concepts, contexts and cultures* (pp. 32–48). London: Routledge.

Epstein, A. S. (2007). *The intentional teacher: Choosing the best strategies for young children's learning*. Washington, D.C.: National Association for the Education of Young Children.

Fleer, M. (2011). 'Conceptual Play' foregrounding imagination and cognition during concept formation in early years education. *Contemporary Issues in Early Childhood, 12*(3), 224–240.

Siraj-Blatchford, J. (2009). Editorial: Education for sustainable development in early childhood. *International Journal of Early Childhood, 41*(2), 9–22.

Trawick-Smith, J. (2012). Teacher-child play interactions to achieve learning outcomes—Risks and opportunities. In R. C. Pianta, W. S. Barnett, L. M. Justice, & S. M. Sheridan (Eds.), *Handbook of early childhood education*. USA: Giuldford Publications.

Wood, E. (2013). *Play, learning and the early childhood curriculum*, (3rd ed.). London: Sage Publications.

About the Authors

Amy Cutter-Mackenzie is Associate Professor in the School of Education in the area of Sustainability, Environment and Education. She is the Director of Research for the School of Education, and the Research Leader of the Sustainability, Environment and Education (SEE) Research Cluster. Amy commenced her career as a primary school teacher in Queensland Australia and later moved into academia after completing her Ph.D. Amy's research is clearly situated in the area of children's and teachers' thinking and experiences in environmental education and sustainability in a range of contexts and spaces (including early childhood education, schools, teacher education, higher education, research and communities). Amy is the Editor of the Australian Journal of Environmental Education (Cambridge University Press) and Consulting Editor for the Journal of Environmental Education, the International Journal of Early Childhood Environmental Education (NAAEE) and International Journal of Environmental and Science Education. She has been recognised nationally for teaching excellence in "leading school-community teaching and learning practices and partnerships to influence, motivate and inspire pre-service education students and schools to engage in environmental education and sustainability" (OLT 2008, 2010).

Susan Edwards is the Deputy Director of the Centre for Early Childhood Futures at Australian Catholic University. She works in the area of early childhood education and specialises in researching aspects of the early childhood curriculum, including play-based learning, teacher thinking, digital technologies and environmental education. Susan has achieved national recognition for teaching excellence in the tertiary sector and has published a number of key texts associated with early childhood education. She is the co-author of Early Childhood Curriculum: *Planning, Assessment and Implementation* published by Cambridge University Press and a co-editor of *Engaging Play* published by Open University Press. Associate Professor Edwards is currently one of two editors for the Asia Pacific Journal of Teacher Education.

Deborah Moore is a Ph.D. candidate in the Centre for Early Childhood Futures at Australian Catholic University. With a background of over 25 years as a Preschool teacher and Preschool Field Officer, Deb was also the inaugural Early Years Sustainability Officer for a local government in Victoria. Deb has worked for many years for Play Australia as one of their Early Childhood Outdoor Play

A. Cutter-Mackenzie et al., *Young Children's Play and Environmental Education in Early Childhood Education*, SpringerBriefs in Education, DOI: 10.1007/978-3-319-03740-0, © The Author(s) 2014

seminar presenters. Deb's Ph.D. research is based around young children's imaginative play places, and her research interests include environmental education for young children and their outdoor play places.

Wendy Boyd is a Lecturer in Early Childhood Education at Southern Cross University (SCU). She was an early childhood educator for 25 years before moving into academic life. Having completed her Ph.D. in 2011, Wendy has collaborated to research and publish in the areas of pre-service teachers' attitudes to child care, early childhood education for sustainability, pre-service teachers' attitudes to mathematics, and teaching in higher education. Her approach to teaching pre-service early childhood teachers has been recognised through the 2011 SCU's Vice Chancellor's Citation Award for Excellence in Teaching. Wendy is an editor for the New Zealand Research in Early Childhood Education Journal.

Author Index

A. Cutter-Mackenzie et al., *Young Children's Play and Environmental Education
in Early Childhood Education*, SpringerBriefs in Education,
DOI: 10.1007/978-3-319-03740-0, © The Author(s) 2014

Subject Index

A. Cutter-Mackenzie et al., *Young Children's Play and Environmental Education in Early Childhood Education*, SpringerBriefs in Education, DOI: 10.1007/978-3-319-03740-0, © The Author(s) 2014

Lightning Source UK Ltd.
Milton Keynes UK
UKOW06f1935230315

248382UK00004B/39/P